Thinking about Thought

*

The Structure of Life and the Meaning of Matter

*

Piero Scaruffi

Volume 4.
Consciousness

*"Intelligence is not about knowing the answers
but about asking the questions"*

*"What we understand is not enough to understand
why we understand it"*

Scaruffi, Piero
Thinking about Thought - Consciousness
All Rights Reserved © 2014 by Piero Scaruffi

ISBN-13: 978-1503362161
ISBN-10: 1503362167

(In the USA only one company is authorized to sell ISBNs: Bowker. And Bowker sells them at an outrageous price when other nations issue ISBNs for free. I consider this as, de facto, one of the most blatant scams in any industry. To protest against this government-sanctioned Bowker ISBN monopoly rip-off, I opted to obtain a free ISBN from Amazon CreateSpace, which will then appear as the publisher of this book, and I encourage all authors and publishers to do the same)

Printed and published in the US

For information: www.scaruffi.com

No part of this book may be reproduced or transmitted in any form or by any means, graphic, electronic or mechanical, including photocopying, recording, taping or by any information storage retrieval system, without the written permission of the author (http://www.scaruffi.com)

Contents

DREAMS 5

EMOTIONS 24

THE HISTORY OF CONSCIOUSNESS 47

CONSCIOUSNESS: THE FACTORY OF ILLUSIONS 70

THE PHYSICS OF CONSCIOUSNESS 118

THE SELF AND FREE WILL: DO WE THINK OR ARE WE THOUGHT? 148

FINALE: WHAT DOES IT ALL MEAN? 172

Preface

By the time you finish reading this book you will be a different person. I am not claiming that this book will change the way you think and act. I am simply referring to the fact that the cells in your body, including the neurons of your brain, are continuously changing. By the time you finish reading this book you will "literally" be a different body and a different brain. Every word that you read is having an effect on the connections between your neurons. And every breath you take is pacing the metabolism of your cells. This book is about what just happened to you.

This volume is one of four in a series titled "Thinking about Thought". See the first volume, "Brain", for the general preface.

DREAMS

The Interpretation of Dreams

Animals sleep, and that's a mystery in itself. When we dream, we are vulnerable: why would Nature select that odd phenomenon? Why do we sleep in the first place? Several hypotheses have been weighed. The evolutionary reason could be that natural selection rewarded individuals who were capable of hiding still at the time of the day when they were most vulnerable. An energetic reason would be that sleep optimizes an individual's energies at the time of the day when it is least efficient to search for food. Others prefer a restorative reason: during sleep the body performs maintenance (muscle growth, tissue repair, protein synthesis, etc. occur during sleep). And there could also be a cognitive reason: during sleep the brain might self-organize.

And then there is an oddity within the oddity of sleeping. The bizarre, irrational nature of dreams, where reality gets warped and laws of nature are turned upside down, and why we remember them at all have puzzled humans since ancient times. Ancient people believed that dreams were the vehicle that the gods used to communicate with mortals. Dreams belonged to the sphere of the supernatural. Dreams were due to external forces. In almost all civilizations people believed that dreams had to be interpreted, i.e. that they had a meaning and that specialists (whether oracles or priests) could figure out that meaning. Aristotle tried in vain to demystify dreams, arguing that their story-line was accidental and meaningless, that they reflected the events of the day, and that they were ultimately caused by imbalances in the body.

However, dreaming is a process that absorbs a lot of energy; therefore, it must serve a biological purpose, possibly an important one.

It is also a curious fact that only warm-blooded animals seem to dream (or, at least, to have REM sleep). It could be an oddity that just happened at the first warm-blooded animals and is still with us today, or it could be that dreaming is a biological necessity for any warm-blooded animal.

In 1900 the Austrian psychoanalyst Sigmund Freud advanced a theory of dreams that stood on the following principles: 1. Dreams are composed of sensory images; and 2. Free associations are evoked in the dreamer's mind by these images. He concluded that dreams rely on memories and that they are assembled by the brain to deliver a meaning. Meaning of dreams are hidden and reflect memories of emotionally meaningful experiences.

The classical world of Psychology was a world in which actions have a motive. Motives are mental states, hosted in our minds and controlled by our minds. Motives express an imbalance in the mind, between desire and reality: action is an attempt to regenerate balance by changing the reality to match our desire.

Freud's conceptual revolution consisted in separating this mechanism of goal-directed action from the awareness of it. Freud suggested that motives are sometimes unconscious, i.e. that we may not be aware of the motives that drive our actions. There is a repertory of motives that our mind, independent of our will, has created over the years, and they participate daily in determining our actions. Our conscious motives, the motives that we can count, represent only a part of our system of desires.

Freud interpreted dreams accordingly. A dream is only apparently meaningless: it is meaningless if interpreted from the point of view of conscious motives. However, the dream is a perfectly logical construction if one also considers the unconscious motives. It appears irrational only because we cannot access those unconscious motives.

Freud's fundamental thesis was hidden (ironically, it was itself unconscious): that all mental life is driven by motives/desires. Freud never justified it and spent the rest of his life analyzing the content of dreams for the purpose of eliciting the unconscious motives (i.e., of "interpreting" the dream), focusing on sexual desires (a prime example of censored motives in his time) and childhood traumas (which somehow he believed were more prone to generate repressed motives).

Freud never even tried to explain the mechanism by which repression of motives operates (by which the unconscious is created, by which some motives are selected over others as undesirable and then such motives are repressed but kept alive and active) and the mechanism by which unconscious motives re-emerge during sleep (by which sleep transforms those repressed motives into a flow of scenes).

Freud's work had an unfortunate consequence: psychiatrists became more interested in the "content" of dreams than in the "form" of dreaming. Psychiatrists kept looking for the "meaning" of dreams, rather than for the process of dreaming. Psychiatrists studying dreams behaved like doctors analyzing symptoms of a disease or of an injury. However, unlike doctors, who knew the anatomy of the body, psychiatrists knew nothing of the neural processes of the brain. This historical accident basically caused dreams to remain outside the sphere of science for five decades. (Freud's impact, incidentally, has always been larger in the arts than in the sciences). It was only in the 1950s that neuroscience, equipped with new tools to record brain waves, began to study dreams.

The Interpretation of Dreaming
Much more important was a finding that remained neglected for almost a century: at the end of the 19th century the British neurologist John

Hughlings Jackson ("On some implications of dissolution of the nervous system", 1882) realized that a loss of a brain function almost always results in a gain in another brain function. Typically what is gained is heightened sensations and emotions. Jackson, virtually a contemporary of Charles Darwin, explained this phenomenon with the view that the brain's functions have different evolutionary ages: newer ones took over older ones, but the older ones are still there, we just don't normally need to use them as the newer ones are more powerful. When we lose one of the newer features, then the older features of the brain regain their importance. Jackson had the powerful intuition: that a single process was responsible for a "balance" of brain states.

One century later the Swiss pharmacologist Alexander Borbély derived a simple law of sleep ("A two process model of sleep", 1982): the timing and intensity of sleep are due to the interaction between the sleep-independent circadian system (that dictates the 24-hour cycle) and a homeostatic mechanism (the less you sleep, the more you need to sleep, and viceversa). The latter expresses a fundamental need for balance between sleep and wakefulness.

An important discovery (probably the one that opened the doors of the neurobiology of dreams) was made by Eugene Aserinsky: dreaming is associated with a brain state of "rapid eye movement" (REM) that recurs regularly during sleep ("Regularly occurring periods of eye mobility and concomitant phenomena", 1953). His advisor, the Moldovan neurologist Nathaniel Kleitman and another student of his, William Dement ("Cyclic variations in EEG during sleep and their relation to eye movements, body motility, and dreaming", 1957), clarified that a brain enters REM sleep 4 or 5 times per night, at approximately 90-minute intervals, and each period lasts about 20 minutes.

Then the French physiologist Michel Jouvet observed that REM sleep is generated in the pontine brain stem, also known as "pons" ("On a stage of rapid cerebral electrical activity in the course of physiological sleep", 1959). In other words, Jouvet localized the trigger zone for REM sleep (and therefore dreaming) in the brain stem.

REM sleep exhibits four main properties:
- A low level of brain activity
- The inhibition of muscle tone
- Waves of excitation from the pons
- Rapid eye movement

The waves of excitation are probably the cause of everything else. The pons sends signals and excites eye muscles (causing rapid eye movement), the midbrain (causing a low level of brain activity), and the thalamus. The thalamus contains structures for visual cognition, auditory cognition, tactile cognition and so forth. The thalamus then excites the cortex. The cortex therefore receives a valid sensory signal from the thalamus and interprets it as if it were coming from the eye, ears, etc.

During REM sleep several areas of the brain are working frantically, and some of them are performing exactly the same job they perform when the brain is awake. The only major difference is that the stimuli they process are now coming from an internal source rather than from the outside world.

REM sleep is pervasive among mammals and birds. Whatever its function is, it has to be the same for rats and humans.

The US neurophysiologist Allan Hobson and the US psychiatrist Robert McCarley argued that, far from being the center of production of dreams as Freud imagined, higher brain functions such as memory and emotion are simply responding to a barrage of stimuli that are generated from the brainstem ("The Brain as A Dream State Generator", 1977). In other words, dreams originate from random brain activity (during REM sleep) which is then interpreted by the forebrain. The brain tries to make sense of what is going on, and what is going on is simply a continuous flow of random "thoughts". Dreams have no meaning.

Sleep

Meanwhile, there was also progress in understanding sleep.

The US psychiatrist Scott Campbell and the Swiss biologist Irene Tobler provided the standard definition of sleep in four steps ("Animal sleep", 1984): 1. A reversible state during which voluntary movements do not occur; 2. Controlled by a circadian clock; 3. Accompanied by an increase in arousal threshold; 4. Controlled by a homeostatic system.

The Romanian neurophysiologist Mircea Steriade discovered the slow oscillations of NREM sleep (~0.5-1 Hz), due to groups of neurons that fire together for a relatively prolonged time (depolarization) and then fall silent for another prolonged time (hyperpolarization), and then resume their synchronized firing ("A novel slow oscillation of neocortical neurons in vivo", 1993).

During the 2000s the US neurologist Clifford Saper identifiet the "sleep switch" in the hypothalamus ("A putative fip-flop switch for control of REM sleep", 2006), and the Taiwanese neurologist Ying-Hui Fu discovered the gene that regulates sleep length in mammals ("The Transcriptional Repressor DEC2 Regulates Sleep Length in Mammals", 2009).

A computational model to explain the function of sleep was proposed by the British computer scientist Geoffrey Hinton ("The wake-sleep algorithm for unsupervised neural networks", 1995). His algorithm requires a dual process of "recognition" and "generation". The top-down connections implement a generative model, the bottom-up connections implement a recognition model. In the "wake" phase the "recognition" proceeds bottom-up from the input to increasingly higher levels of representation. At the same time, this recognition model that is being created tweaks the parameters of the generative model so as to maximize the probability that the latter will reconstruct the input. In the "sleep" phase the "generation" proceeds top-down from the representation while the recognition model is being tweaked so as to maximize the probability of reconstructing the activity that is going on. The wake-sleep cycle helps the neural network learn in a way that a wake-only process would not.

Dreams Are Made Of This

There are three main categories of explanation for dreams. The simplest explanation is that dreams are just an evolutionary accident. By accident we have five fingers rather than four. By accident we dream while we sleep. Another explanation is that they are fossils of a previous form of mind, accidental remnants of previous brain processes. Yet another explanation is that they are a window on some kind of processing that goes on in the brain while we sleep.

Imagine that somebody is filing a lot of newspaper clippings into folders and that you are standing in front of him: you will see a rapid sequence of titles flashing in front of your eyes. While you understand each of them, the flow of titles is cryptic: it may form, by mere chance, stories, but stories that you cannot understand. In reality the sequence of titles is not random, because the person who is filing the titles is following a method (for example, they are filed in chronological order, or in order of importance, or by subject matter). It is just that you are only a spectator of the process, trying to make sense of the output of that process. This could be exactly what is happening to our consciousness while we are sleeping. The brain is rapidly processing a huge amount of information in whatever order and our consciousness sees flashes of the bits that are being processed. These bits seem to compose stories of their own, and no wonder that the stories look weird if not indecipherable.

This third hypothesis is consistent with the behavior of the brain. The brain, far from being asleep, is very active during sleep. Most nerve cells in the brain fire all the time, whether we are awake or asleep.

The process that takes place during dreams is most likely about remembering and forgetting, i.e. REM sleep is important for consolidating long-term memories.

An important clue is that the brain is not only very active during sleep, but sleep states and awake states are quite similar.

The Genetic Meaning Of Dreams

Jouvet was also a pioneer of the theory that dreams have a function. To derive crucial action patterns from the genetic program of the individual. REM sleep may provide a means to combine genetic instructions with experience. The combination of sleep and dreaming may just be survival strategy.

His theory of "iterative genetic programming" ("Paradoxical sleep and the nature-nurture controversy", 1980) is based on the observation that, evolutionarily speaking, there seems to be an inverse relationship between post-natal neurogenesis (the brain of cold-blooded animals continues to grow throughout their lifetime, whereas the brain of warm-blooded animals is largely shaped at birth) and REM sleep (widespread among warm-blooded animals). Jouvet concludes that REM sleep could play the same role that neurogenesis does: continuously recreate the self. Within the debate of nurture and nature, Jouvet side with those who think that we don't simply inherit from our parents some somatic (bodily) traits, but that we also inherit traits of a personality. The events of a day tend to change that personality. At night the "original" personality gets reinforced during dreams, whose goal is to erase knowledge that is harmful to that personality and to keep what is useful.

More than Freud's pathological theory of dreaming, this resembles the theory of the Swiss psychologist Carl Jung, that dreams reflect the "collective unconscious", a shared repertory of archaic experience represented by "archetypes" which spontaneously emerge in all minds. One only has to adapt Jung's thought to genetics in order to obtain Jouvet's theory. The universal archetypes envisioned by Jung could represent a predisposition of all human brains to create some myths rather than others, just like, according to Chomsky, all human brains inherit a predisposition towards acquiring language.

The Adaptive Value Of Dreams

The US neurobiologist Jonathan Winson expressed this concept in a more general way: dreams represent "practice sessions" in which animals (not only humans) refine their survival skills.

REM sleep helps the brain to "remember" important facts without having to add cortical tissues. During REM sleep the brain (specifically, the hippocampus) processes information that accumulated during the day. In particular, during REM sleep the brain relates recent memories to old memories, and derives "tips" for future behavior. Dreams are a window on this "off-line processing" of information.

The neocortex processes sensory input and sends it to the hippocampus, which acts as a gateway to the limbic system. The limbic system mediates between sensory input and motor output.

Winson used three main clues to reach his conclusions.

1. At birth, the hippocampus is needed to retrieve information stored in long-term memory, but, after about three years, the brain somehow learns how to access directly such information.

2. During REM sleep, the time when we dream, the neocortex is working normally, except that movement in the body is inhibited.

3. Most mammals, except for primates, exhibit a theta rhythm in the hippocampus (about six times per second) on only two occasions: whenever they perform survival-critical behavior, and during REM sleep. Therefore, REM sleep must be involved in survival-critical behavior.

Early mammals had to perform all their "reasoning" on the spot ("on-line"). In particular they had to integrate new information (sensory data) with old information (memories) immediately to work out their strategies. Winson speculates that at some point in evolution brains invented a way to "postpone" processing sensory information by taking advantage of the hippocampus: REM sleep. Theta rhythm is the pace at which that ("off-line") processing is carried out. Instead of taking input from the sensory system, the brain takes input from memory. Instead of directing action, the brain inhibits movement. But the kind of processing during REM sleep is the same as during the awake state. Winson speculates that this off-line processing is merging new information with old memories to produce strategies for future behavior.

Theta rhythm disappeared in primates, but REM sleep remained as a fundamental process of brains. In humans, therefore, REM sleep, i.e. dreams, corresponds to an off-line process of integration of old information with new information.

Dreaming is an accidental feature that lets us "see" some of the processing, although only some: a dream is not a story but a more or less blind processing of the day's experience.

Winson goes as far as to suggest that all long-term memory may be constructed through this off-line process (i.e., during REM sleep): the hippocampus would process the day's events and store important information in long-term memory.

There is a biologically relevant reason to dream: a dream is an ordered processing of memory that interprets experience that is precious for survival. Dreaming is essential to learning.

Winson relates this off-line process that operates during sleep with Freud's subconscious. Freud was right that dreams are the bridge between the conscious and the unconscious, although that bridge is of a different nature. The Freudian "subconscious" becomes the phylogenetically-ancient mechanism involving REM sleep, in which memories and strategies are formed in the cortex.

Similarly, Hobson thinks that the ultimate purpose of dreams is to populate long-term memory, to help us learn. We dream hypothetical situations so that we will be prepared to face real situations of the same kind. When a situation occurs during the day, it has probably already been played at least once at night in our dreams, and we know what to expect. By dreaming, we train our brain: dreams are mental gymnastics. It's like saying that, in order to see something, we must first create the vision of that something in our mind.

In a sense, we dream what is worth remembering.

We Dream to Remember and to Forget

By contrast, the British biologist Francis Crick, revisiting an old pre-Freudian theory by the German physician W Robert, proposed that the function of dreams is to "clear the circuits" of the brain, otherwise there would not be enough space to register each day's events ("The function of dream sleep", 1983).

The brain, in the face of huge daily sensory stimulation, must:
- understand what matters
- understand what does not matter
- remember what will still matter
- forget what will never matter again

Dreams help eliminate useless memories. Therefore, according to Crick, we dream what is worth forgetting. If we didn't, the neural network of the brain would quickly become incapable of performing its tasks. "We dream in order to forget".

Robert argued that harmful or useless thoughts had to be eliminated from the mind lest they caused psychological imbalances. The dream shows us that process as it happens. Dreaming is a process to improve the mind.

What is still missing is the physical link between dreams and genome. Neurotransmitters (such as animenes and cholines) act on the surface (the membrane) of the cell, whereas genes lie in the center (the nucleus) of the cell. Messenger molecules transfer information from the membrane to the nucleus and viceversa. Allan Hobson has hypothesized that neurotransmitters may interact with messenger molecules and therefore affect the work of genes.

If during sleep the brain is consolidating memories that have been acquired during the day, then dreaming, far from being an eccentric manifestation of irrationality, is at the core of human cognition.

Dreams as a Chemical System

Hobson thinks that, whether asleep or awake, the brain always does pretty much the same thing. The dreaming brain employs the same systems and processes as the brain when awake, except that those processes are not activated by stimuli from the outside world; that the outcome of those processes does not result in (significant) body movements; and that self-awareness and memory are dormant. The input, the output, the processor and the working space of the brain while awake are replaced by something else; but the "software" is the same.

What makes a difference is the neurotransmitters that travel through the brain. What differs between awake and asleep states is very small, but enough to alter dramatically the outcome: during sleep the brain is bombarded by erratic pulses from the brain stem and flooded with nervous system chemicals of a different sort.

Neurotransmitters make brain circuits more or less sensitive. Aminergic neurotransmitters originate in the brain stem and terminate in the amygdala. Cholinergic neurotransmitters originate in the basal forebrain and terminate in the cortex. During waking states, the brain is controlled by the aminergic neurotransmitters, made of molecules called "amines". During sleep, the brain is controlled by the cholinergic neurotransmitters, made of a molecule called "acetylcholine". Cholinergic chemicals free the system used for cognition and behavior. They paralyze the body by sending pulses to the spinal chord, even though motor neurons are always in motion.

Hobson's idea is that wake and sleep are two different chemical systems hosted in the same "processor".

These two chemical systems are in dynamic equilibrium: if one retracts, the other advances. This means that our consciousness can fluctuate between two extremes, in which either of the chemical systems totally prevails (neither is ever completely absent). This also means that the brain states of wake and sleep are only two extremes, between which there exists a continuum of aminergic-cholinergic interactions, and therefore a continuum of brain states. This system can be said to control the brain. It resides in the brain stem and from there it can easily control both the lower brain (senses and movement) and the upper brain (feelings and thought).

When it doesn't work properly, when the balance of chemicals is altered, mental diseases like delirium occur. It is not surprising that diseases such as delirium are so similar to dreams: they are driven by exactly the same phenomenon.

Hobson claims that the brain is in awake, dream or (non-REM) sleep mode depending on whether amines are prevailing, cholines are prevailing or amines and cholines are "deadlocked".

Three factors account for the brain behavior at any time: activation energy (amount of electrical activity), information source (internal or external) and chemical system (amines or cholines).

When activation energy is high, the information source is external and the mode is aminergic: the brain is awake. As activation energy decreases, the external information source fades away and amines and cholines balance each other: the brain falls asleep. When activation energy is high, the information source is internal and the mode is cholinergic: the brain is dreaming. During an hallucination: activation energy is high, the information source is internal and the mode is aminergic. In a coma: activation energy is low, the information source is internal and the mode is cholinergic.

The extremes are rare and usually traumatic. Normally, both external and internal sources contribute to cognitive life, and both amines and cholines contribute to the brain state.

The interplay of external and internal sources means that our perceptions are always mediated by our memory. Hobson thinks that our brains do not merely react to stimuli: they also "anticipate". The internal source tells us what to expect next, and thus helps us cope with the external source. Emotions are, in a sense, a measure of how well the internal source matches the external source: anxiety is caused by a major mismatch, whereas contentedness is a sign of matching sources.

When we dream, the spinal cord is paralyzed and the senses are disconnected. This is because of the cholinergic neurotransmitters that come from the brain stem.

Hobson believes that sleep has the function to reinforce and reorganize memory: ultimately, to advance them from short-term memory to long-term memory. Amines are necessary for recording an experience, while cholines consolidate memory. Again, it looks like during REM sleep memory is consolidated.

The aminergic system is responsible for attention, focus, awareness. The cholinergic system is responsible for the opposite process: focus on nothing, but scan everything.

As for the content of dreams, Hobson thinks that they reflect a biological need to keep track of place, person (friend, foe or mate) and time. He draws this conclusion from considerations about what is typical (and bizarre) of dreams: disruptions in orientation.

The bottom line is that dreams are meaningful: the mind makes a synthetic effort to provide meaning to the signals that are generated internally (during a dream, memory is even "hypermnesic", i.e. is intensified). Wishes are not the cause of the dreaming process, although, once dreaming has been started by the brain stem, wishes may be incorporated in the dream. Therefore, Hobson thinks that dreams need not be interpreted: their meaning is transparent. Or, equivalently, dreams must be interpreted in the realm of neurophysiology, not psychology.

The interplay between the aminergic and the cholinergic systems may be responsible for all conscious phenomena (for Hobson, dreams are as conscious as thinking) and ultimately for consciousness itself. After all, conscious states fluctuate continuously between waking and dreaming.

Dreams, far from being subjective, are "impersonal necessities forced on brain by nature".

The Origin of Dreaming

The US psychiatrist Fred Snyder advanced the hypothesis that, from an evolutionary perspective, REM sleep came first and dreams came later. First bodies developed the brain state of REM sleep, which was retained because it had a useful function for survival (for example, because it kept the brain alert and ready to react to emergencies even during sleep), and then dreams were engrafted upon REM sleep. REM sleep was available and was used to host dreams. Dreaming evolved after a physical feature made them possible, just like language evolved after an anatomical apparatus that was born for whatever other reason. Dreaming, just like language, is an "epiphenomenon". The real purpose of REM sleep was to act as a "sentinel".

The psychologist Anthony Stevens has provided a practical explanation for why some animals started dreaming: dreaming emerged when oviparous animals evolved into viviparous animals. By dreaming, the brain could augment its performance with some "off-line" processing. This made it possible to limit the size of the brain while leaving brain activity free to grow. Brains, and thus heads, would remain small enough to pass through the maternal pelvis.

In Winson's scenario, dreams helped us survive a long time before our mind was capable of providing any help at all. And dreams, unlike higher consciousness, are likely to be common to many species.

The mind could well be an evolution of dreaming, which happened in humans and not in other species. First the brain started dreaming, and then dreams took over the brain and became the mind, which could be viewed as a continuous dream of the universe that we inhabit.

This hypothetical history of the mind does not differ too much from the one in which the mind was created by "memes" (concepts that spread from mind to mind). The relationship between memes and dreams is intuitive, and the psychologist Joseph Campbell indirectly summarized it with his celebrated aphorism that "a myth is a public dream, a dream is a private myth".

An Evolutionary Accident

In contrast to Jouvet, Hobson and Winson, the US philosopher Owen Flanagan thinks that both sleep and consciousness are products of evolution, but consciousness during sleep (dreaming) is merely an accident of nature, a side effect of the two. Both consciousness and sleep have a clear biological function, but dreams don't.

During sleep the brain stocks up neurotransmitters that will be used the next day. By accident, pulses that originate from this stockpiling chore (coming from the brain stem) also reactivate more or less random parts of memory.

Unaware that the body is actually sleeping, the sensory circuits of the cerebral cortex process these signals as if they were coming from outside and produce a chaotic flow of sensations. Thus we dream.

Dreams are just the noise the brain makes while working overnight.

If Flanagan is correct, dreams are meaningless and pointless.

Of course, indirectly, dreams tell us something about how our mind works, because, after all, whatever we perceive while we dream is the product of what is in our memory and of how our cerebral cortex processes those memories. But the usefulness of the dream-narrative is really limited to an almost "diagnostic" purpose. As our cerebral cortex tries to make sense of that chaotic input, we can learn something about its cognitive functioning, just like by running the engine of a car when it is not moving we can learn something about a noise it makes on the freeway (but the engine running while the car is not moving has no hidden meaning).

Of How Real Dreams Are and How Dreamy Reality Is

The experience of a dream may feel so utterly bizarre for today's mind, but we have to go back millions of years to realize that it is probably far less bizarre than it appears to us today. It is likely that, millions of years ago, our waking life was not too different from our dreaming life. Consciousness in dreams is a series of flashes that are fragmented and very emotional. It is likely that waking consciousness had exactly the same character: mostly nothing would happen to our consciousness (no thinking, no emotions, just mechanic, instinctive behavior) but situations would present themselves suddenly that would arouse strong feelings and require immediate action. Our waking life "was" a series of emotionally charged flashes, just like dreams. The difference between being awake and dreaming was only the body movement. As we rehearsed the day's events during dreams, we would feel that the sensations are perfectly normal.

Today our consciousness has acquired a different profile: it has evolved to a more or less smooth flow of thoughts, in which strong emotions don't normally figure prominently. We think when we are commuting on a bus or while we are shopping in the mall, and the most violent emotion is being upset about the price of a shirt or suddenly realizing we just missed our stop. They are peanuts compared with the emotion of being attacked by a tiger or of being drawn by strong currents towards the waterfall. Our waking consciousness has changed and dreams have remained the same. The brain is still processing off-line, during sleep, our day's events with the same cerebral circuits that we had millions of years ago, and therefore it is still generating the same flow of emotionally-charged flashes of reality. When the brain is awake, reality does not impinge on those circuits in the same way it did in the hostile, primitive environment of million of years ago. The world we live in is, by and large, friendly (free of mortal foes and natural catastrophes). But when danger does appear (a mortal foe or a natural catastrophe), then our waking life becomes just like a dream: "it was a nightmare", "it didn't feel real", etc. In those rare and unfortunate circumstances (that hopefully most of us will never experience) our waking life feels just like a dream: flashes of reality, violent emotions, apparent incoherence of events, etc.

Because of the society that we have built and the way we have tamed and harnessed nature's unpredictability through civilization, our brain does not receive the sudden and violent stimuli it used to. This is what makes most of the difference between being awake and dreaming. It is not just a different functioning of the brain: it is a different functioning of the world around us.

Dreams are Energetic

Most of the cells in the brain are actually not neurons but astrocytes or astroglia. These cells provide many functions to the nervous system and, in particular, store and release glycogen. Ramon y Cajal had already speculated that sleep might have to do with these cells. The Swedish neurologists Holger Hyden and Paul Lange ("Rhythmic enzyme changes in neurons and glia during sleep and wakefulness", 1965) discovered that the level of enzyme activity is higher in neurons during sleep and slower during wakefulness, and viceversa in glia. It took three decades but eventually the US biologists Craig Heller and Joel Benington formulated the theory that the main role of sleep is to restore the brain's energy. Adenosine is a nitrogen-based building block of adenosine triphosphate, or ATP, the energy-rich engine of the cell, stored as glycogen in the brain. During wakefulness ATP is consumed. Whenever the cell needs energy because the ATP reserves are being depleted, adenosine is released by brain cells. In other words, the processes of perception, thought and action use up the glycogen stores in the brain. Adenosine is an inhibitory neurotransmitter. Eventually adenosine reaches a level that inhibits signal propagation and triggers non-REM sleep. Basically, adenosine triggers a chemical process that signals the depleted cells to enter a state of rest. In this state the brain is less active, consuming much less glycogen and energy can be replenished with new glycogen. Glycogen stores take time to replenish. The paradox (as Michel Jouvet called it) is that dreams use up a lot of this energy.

This homeostatic process works independently from the circadian clock. The US psychiatrists Dale Edgar and William Dement ("Evidence for Opponent Processes in Sleep/Wake Regulation", 1992) showed that the circadian clock, in fact, keeps us awake during the day, even when we get tired and technically we should be falling asleep to recharge our energy storage. Other animals don't have such opposing clocks and sleep whenever they need energy restoration.

The Colombian neurophysiologist Rodolfo Llinas tried to explain the paradox of dreaming away by hinting that dreaming could be a fundamental state of the brain and wakefulness could simply be an offshoot of dreaming ("Of dreaming and wakefulness", 1991).

NREM Sleep

The South African psychiatrist Mark Solms argued that the identification of dreaming and REM is incorrect, because, he believes, dreaming is possible without REM (Non-REM dreaming or NREM dreaming). He also claimed to have located the origins of dreaming in the ventral tegmental area of the midbrain, and not in the pons (as Jouvet and Hobson believed). Therefore Solms argued that the dopaminergic system (the one originating in the ventral tegmental area) is the neurochemical basis for dreaming.

Hobson believed that dreaming has its origins in the same region of the pons that generates REM sleep. Solms believes that only REM originates in the pons, i.e. that dreams and REM are physically controlled by two different regions of the brain.

The Canadian psychiatrist Jie Zhang argued that dreaming and REM sleep must be controlled by different parts of the brain ("Memory process and the function of sleep", 2004). The two kinds of sleep serve two different purposes. The day's memory are stored in a temporary memory. This working memory consists of two subsidiary systems: the conscious and the non-conscious subsidiary systems. NREM sleep is for processing the declarative (conscious) memory, and REM sleep is for processing the procedural (nonconscious) memory. When the unconscious procedural memory is transferred from the temporary memory to the long-term memory during REM sleep, the conscious declarative memory enters Hobson's "continual-activation" mode to interpret the memory stream. Dreaming is only an epiphenomenon of the conscious subsidiary system of working memory.

Selectionism
William Calvin introduced the idea that the brain works all the time just like Nature works all the time at producing and selecting species. The brain is a small ecosystem of "thoughts" that are continuously produced in countless variations and then selected depending on which one suits the current conditions.

In this sense we dream all the time. When we are awake, the rational part of the brain selects the "dreams" that make sense and disposes of the ones that don't make sense. When we are asleep, that "rational" center is not working (its inhibitory power is relaxed) and therefore our mind is flooded with random dreams. The difference between a dream and a thought is basically just that the dream has not been certified by the rational center.

The US neuropsychologist Patrick McNamara resurrected the Darwinian theory of dreaming embraced a century earlier by Henri Bergson and William James that predated this idea.

The US psychologist Mark Blechner proposed "oneiric Darwinism": during dreams existing ideas can mutate into new ideas, which are then either retained or discarded depending on how well they match reality.

The Collective Memory of Myths
Influenced by Carl Jung, in the 1940s the US anthropologist Joseph Campbell argued that a few themes are ubiquitous in myths around the world. Myths recur in different civilizations and evolve from one civilization to the next one. There is a continuum of myth. At the origin of a myth there is an archetype, which works as a "memory deposit". Mythology seems to be a system of entities conceived to mirror the human condition. The rites are "physical formulas" in human flesh, unlike mathematical formulas that are written in symbols, but they are also formulas that describe natural laws of the universe.

Myth appears to be a system which organizes knowledge about the environment and passes it on to others, in particular to future generations. The reason this system works is that it somehow takes advantage of the way the human brain works. A myth is so constructed that, once inserted in a human brain, it will provoke the desired reaction. It does not require thinking. In a sense, it "tells" you what to do, without having to prove that it is the right thing to do. It shows you the consequences, or the rewards, so you are prepared for them; or it shows you the dangers and so it saves you from experiencing them in real life. For example, when the Sumerian city yields the myth of the city of god what matters is not the historical record but the subliminal message: build such a city! The creator of the myth must craft the myth in such a way that it will trick the brain into wanting to achieve a goal. The "creator", naturally, is not one specific author, but rather the chain of humans who use and adapt the myth to their conditions. The myth evolves over many generations and eventually gets "designed" to perform its task in a very efficient way, just like any organ of the body.

Campbell calls myth "the spiritual resources of prehistoric man" and insists on the "spiritual unity" of the human race: the spiritual history of the human race has unfolded roughly in the same way everywhere.

Campbell also implies that myth, just like language and just like genes, obeys a grammar. Just like language and just like genes, myths have evolved from more primitive myths. Just like language and just like genes, myths are universal, shared by all humans. Just like language and just like genes, myth tells the story of mankind, as it follows the spread of races in the continents and their adaptation to new natural pressures.

Campbell's viewpoint contrasts with that of the British anthropologist James Frazer, who at the end of the 19th century claimed that the similarity of myth is due to similar causes operating on similar brains in different places and times. But there may be a bit of truth in both views.

A Neuroscience of Myths

Noam Chomsky argued that brains are "pre-wired" for learning language. Circumstances will determine which particular language you will learn, but the reason you will learn a language is that your brain contains a "universal grammar" that is genetically prescribed. The structure of the human brain forces the brain to learn to speak.

Dreams seem to be the antichamber to learning, whereby the brain processes experience and decides what has to be stored for future use. Again, this depends on the genetic programming of the brain. What is learned depends on the structure of the brain, which in turn depends on the genetic program that created the brain, which in turn was determined over millions of years by evolution.

Basically, it looks like brains learn what they have been programmed to learn by evolution. Human brains have been programmed by evolution to learn some things rather than others.

If this is true, then myths could have a simple explanation: they are the simple ideas that the structure of our brain can accommodate.

The universal archetypes envisioned by Jung and Campbell could well be predispositions by all human brains to create some myths rather than others, just like, according to Chomsky, all human brains inherit a predisposition towards acquiring language.

If myths arise in the mind because the brain has been programmed to create them, then, to some extent, we think what we have been programmed to think.

Down this path of genetic determination of our beliefs, one could wonder if genetic differences between the races can account for slightly different behavior. Do Italians speak Italian and French speak French because their brains developed in slightly different ways over thousands of years of independent evolution? Did Arabs develop Islam and Europeans Christianity because their brains are slightly different, and the myths they can create are slight variations of the myth of god?

How plastic is our mind, and how strong is the hand of fate?

A History of Concepts

One could argue that ancient gods simply represent concepts. As concepts were forming in human minds, human minds expressed them as concepts. And their interaction yielded religions.

Ancient gods represented qualities, mountains, rivers, forces of nature, emotions. Gods were vehicles for natural forces to express themselves. Gods were created for each new phenomenon or discovery. Ancient religions were systems of concepts: they classified and organized concepts through a network of legends (symbolic narratives) and a series of rites (symbolic actions) in a manner not too different from how Marvin Minsky's "frame" organizes knowledge. A legend expressed the function of the force in nature and in relation to other forces. A rite expressed the attributes of the force.

They were "local": each culture developed different sets based on local circumstances.

As lower concepts gave rise to higher, more complex concepts, old gods were replaced by new gods (e.g., the god of thunder or of the Nile was replaced by the god of fertility or of harvest). New gods were created as the mind progressed from purely narrative/literal concepts to more and more abstract concepts.

At some point there arose the concept of all concepts, the concept of Nature itself, the concept of the supreme power and of everything that exists, of the force that is behind all other forces. Thus monotheistic religion was born.

God became not the vehicle for a force but "the" force itself.

Religion is, ultimately, a way to pass culture (a system of concepts) on to future generations. People of different religions have not only different rites (physical lives) but also different mental lives, because they think according to different conceptual systems. There is no one-to-one correspondence between Roman gods and Indian gods, and that explains why ancient Rome and ancient India were completely different societies: they had completely different concepts, i.e. people thought in completely different manners.

In that sense, monotheistic religion represents a major leap forward in cognitive skills. Just like the zero enabled higher mathematics and the Roman arch enabled taller buildings, so the single-god religion enabled higher thinking.

Child's Play

Why do children play? Isn't that also a way of rehearsing real-life situations based on the genetic repertory? Doesn't play also have the irrational quality of dreams? Isn't playing a way to accelerate the same kind of learning that occurs during dreaming? Could it be that at the beginning of our life everything is but a dream, and then slowly reality takes over (thanks to social interaction) and dreams are relegated to sleep?

Joking

What have joking and dreaming in common? Apparently nothing, but both belong to the category of acts that do not seem to have a useful function. Like dreaming, joking seems to be a pointless waste of energies. Like dreaming, joking is some kind of playing with our experience. Like dreaming, joking is a process of rearranging our experience in a somewhat irrational way. Like dreams, jokes do not necessarily require linguistic skills, but normally occur in a linguistic context. More than dreams, jokes seem to have developed in humans to a level far more sophisticated than in any other species. We see animals play and laugh, but the gap between a comedian and two lion cubs wrestling in the grass is enormous.

While there may be no biological evidence to support the idea that jokes have a specific function for our learning and survival, one wonders why we enjoy so much making them. Woody Allen once said that comedy is tragedy plus time: when something tragic occurs, it is inappropriate to make fun of it, but months or years later it may be perfectly appropriate. If I trip on something and break my leg, I am in no mood to hear a joke about it, but it is more than likely that years later somebody will mock me on this subject. Jokes refer to past experience, and usually refer to tragic experience. If not tragic, then significant in some way. The point is that, indirectly, jokes help us to learn and to remember.

Jokes help us rehearse survival techniques in the environment. Jokes help us prepare for reality. Jokes tell us which situations we should avoid at all costs. Jokes, like dreams, are a brain's way to train itself without risking its life.

Further Reading
Blechner, Mark: THE DREAM FRONTIER (Analytic Press, 2001)
Calvin, William: THE CEREBRAL CODE (MIT Press, 1996)
Campbell, Joseph: THE MASKS OF GODS (Viking, 1959)
Flanagan, Owen: DREAMING SOULS (Oxford Univ Press, 1999)
Hobson, Allan: THE DREAMING BRAIN (Basic, 1989)
Hobson, Allan: THE CHEMISTRY OF CONSCIOUS STATES (Little & Brown, 1994)
Jouvet, Michel: LE SOMMEIL ET LE REVE/ THE PARADOX OF SLEEP (Jacob, 1993)
Jung, Carl: THE ARCHETYPES OF THE COLLECTIVE UNCONSCIOUS (1936)
Marcos, Frank: THE MYSTERY OF SLEEP FUNCTION (2006)
McNamara, Patrick: MIND AND VARIABILITY: MENTAL DARWINISM, MEMORY, AND SELF (Praeger, 1999)
Pace-Schott, Edward: SLEEP AND DREAMING: SCIENTIFIC ADVANCES AND RECONSIDERATIONS (Cambridge Univ Press, 2003)
Robert, W: DREAMS UNDERSTOOD AS NATURAL NECESSITY (1886)
Snyder, Frederick: EXPERIMENTAL STUDIES OF DREAMING (Random House, 1967)
Solms, Mark: THE NEUROPSYCHOLOGY OF DREAMS (Erlbaum, 1997)
Winson, Jonathan: BRAIN AND PSYCHE (Anchor Press, 1985)

EMOTIONS

Feeling and Thinking

Emotion appears to be a key component in the behavior of conscious beings. To some extent, consciousness "is" emotion. There is probably no recollection, no thinking and no planning that occurs without feeling emotions. As we think, we are either happy or sad or afraid or something else. There is hardly a moment in our conscious life when we are not feeling an emotion. William James conceived mental life as a "stream of consciousness", each state of consciousness possessing both a cognitive aspect and a feeling aspect. James also speculated that emotion is only a secondary feeling: a situation first generates a primary feeling, which is physiological in nature (for example, shivering) and then this physiological event triggers a corresponding emotion. For example, our eyes are wide open and we stop breathing when we witness a car accident, and then these physiological facts trigger the emotion of anxiety.

Whether all of consciousness is just emotion or whether emotion is a parallel, complementary facility of the mind, is debatable. But it can be argued that we would not consider conscious a being that cannot feel emotions, no matter how intelligent it is and no matter how much its body resembles ours.

On the other hand, we ascribe emotions to beings that we do not consider as "conscious": dogs, birds, even fish and tarantulas. Are the intensities of their emotions (of fear, for example) as strong as ours, regardless of whether their level of self-awareness is comparable to ours? Is emotion a more primitive form of consciousness, that in humans developed into full-fledged self-awareness? Is emotion an organ, just like feet and tails, that a species may or may not have, but which has no direct impact on consciousness?

Emotions were traditionally neglected by scientists researching the mind, as if they were a secondary aspect (or simply a malfunction) of the brain activity. The fact is surprising because emotions have so much to do with our being "aware", with differentiating intelligent life from dead matter and non-intelligent life. While the relationship between "feeling" and "thinking" is still unclear, it is generally agreed that all beings who think also feel. That makes feelings central to an understanding of thinking.

The relationship between emotion and cognition, in particular, was first thoroughly explored in 1980 by the US psychologist Robert Zajonc, who emphasized how they are largely independent, and, contrary to popular belief, emotion tend to prevail over cognition in decision making.

That emotions may not be so peripheral a notion as the scant literature on them would imply is a fact suspected since ancient times, but only recently science has focused on their function, their evolution and their behavior. In other words: how did the ability to feel emotions originate, why did it originate and how does it influence our mind's overall functioning?

Emotions as Survival Instinct

The answers can be summarized, once again, as: emotions are a product of evolution, they exist because they favor our species in natural selection. What emotions seem to do is help us make fast decisions in crucial situations. Emotions are inferential short-cuts. If I am afraid of a situation, it means that it is dangerous: the emotion of fear has already helped me make up my mind about how to approach that situation. If I were not capable of fear, my brain would have to analyze the situation, infer logically what is good and what is bad about it for me, and finally draw a conclusion. By that time, it may be too late. Fear helps us to act faster than if we used our logical faculties.

This is reflected in the way emotions are generated. The central processor for emotions is the brain structure called "amygdala" The thalamus normally connects senses to the cortex and the cortex to the muscles. But the amygdala provides a much faster shortcut for decision making: the route from senses to amygdala to thalamus to muscles is much faster than going through the cortex.

However, there is still little evidence for how emotions are implemented in the brain. The Estonian-born psychologist Jaak Panksepp identified seven regions of the mammalian brain corresponding to seven fundamental emotions: seeking, fear, panic, rage, play, care and lust. This model has been widely used, but its correspondence to reality is debatable.

The Emotional Brain

The US neurophysiologist Joseph Ledoux believes that there exist specialized brain circuits (neural maps) for each emotion. Such circuits create as many "shortcuts" to decision making.

The amygdala stores emotional memories, or, more precisely, the amygdala is the place where the brain decides whether to react or act. When we find ourselves in danger, first we "react" (we unconsciously apply one of the patterns of behavior that evolution has added to our repertory of survival strategies) and then we "act" (we make a conscious decision). Unconscious reaction occurs when information flows from the lateral amygdala to the central amygdala (and results in the typical behavior of fear, such as blood pumping, heart thumping, frantic breathing, sweating, pupil dilation). Conscious action occurs when information flows from the lateral amygdala to the basal amygdala (and results in retrieving memories from the neocortex). The transition from unconscious reacting to conscious acting is therefore a "switching" of the flow of information from the lateral amygdala to either the central amygdala or the basal amygdala. The pathway of conscious action is probably unique to humans, whereas the pathway of unconscious reaction is largely shared by all mammals. There is continuity between the emotional brain system of ancient mammals and humans.

Even before birth, the amygdala of a baby memorizes fear states (first of the mother, then of itself). At the same time, the amygdala retrieves and reenacts a fear state whenever a known context recurs. The amygdala stops performing the memory task when the child is about five years old, but continues to work as a template for fear states for the rest of the adult life. Unconscious fear memories are forever.

A critical finding was that information reaches the amygdala before the cortex. In fact, some stimuli may never reach the cortex at all. That is why we may react to a situation without even realizing what we are doing until well after we have done it.

Ledoux believes that emotion (the unconscious reaction) and cognition (the conscious action) are separate but interacting mental functions mediated by separate but interacting brain systems.

Ledoux believes that emotions are a prerequisite to consciousness, but consciousness requires more. It is in the working short-term memory that one becomes aware of one's own emotions. This area is localized in the lateral prefrontal cortex, which only exists in primates and is particularly large in humans.

Emotions as Communication

Emotions are also the fastest way that we can communicate with members of our group, another activity that is critical to survival. The US psychiatrist Allan Hobson thinks that emotions are signals between animals of the same species that communicate one's brain state to another.

Emotions may predate language itself as a form of communication. The US anthropologist Ray Birdwhistell coined a term, "kinesics", for paralinguistic body communication, such as facial expression ("Introduction to Kinesics", 1952). Birdwhistell thought that all movements of the body obey some kind of meaning. Non-verbal behavior has its own grammar, with a "kineme" being the kinesic equivalent of the phoneme.

Kinesics is about emotions. It may well be that body communication existed before language was invented, and that it was the main form of communication.

Facial expression is inevitable like language and universal like language.

Emotion as Rationality

Zimbabwe-born mathematician Aaron Sloman ("Towards A Grammar Of Emotions", 1982), extending original ideas by Herbert Simon ("Emotional And Motivational Controls Of Cognition", 1967), reduced the argument about emotions to simple mathematical terms. Let us look at an "agent" (whether a human, an animal, a robot or a piece of software), which is limited and intelligent and must act in a complex environment. A "complex" environment may well include a very large number of factors. In fact, it may be made of an "infinite" number of factors, if one counts every little detail that may have an influence. Our cognitive agent, which is limited, would never reach a conclusion on what to do if it blindly analyzed all factors. Therefore, in order to survive (to move at all, actually) it must be endowed with mechanisms that cause emotions. In other words, emotions are the result of constraints by the environment on the action of the intelligent being.

An emotional state is created by a situation, through some chemical reaction in the nervous system. A cognitive state is created by a number of situations and by a thinking process that relates those situations and draws some kind of conclusion. The relationship between emotional states and cognitive states is reduced by Sloman to the need to draw conclusions when cognition would face a combinatorial explosion of possible reasoning threads. Emotions emerge when a cognitive agent needs to make survival decisions in a complex environment.

The Canadian philosopher Ronald de Sousa expresses this fact in a different way: emotions play the same role as perceptions, i.e. they contribute to create beliefs and desires. Beliefs and desires are necessary elements of any logical system: one attempts to satisfy desires by acting in the environment according to one's beliefs.

DeSousa believes that emotions are learned like a language. And, like any language, they have their own grammar, i.e. their syntax and semantics (an idea also advanced by Sloman). Just like the meaning of words ultimately derives from the sentences in which they can be used, the semantics of emotions derives from the scenarios in terms of which they have been learned. Emotions can therefore be studied in a formal way just like any other language. The complementarity between reason and emotion becomes what he calls "axiological rationality", yet another way to express the fact that emotions determine what is salient, i.e. can restrict the combinatorial possibilities that reason has to face in the real world.

Emotion as Homeostasis

The US psychologist Ross Buck studied the advantage of emotions in communication between humans. Communication of emotions is a biologically shared signal system. It was created over millions of years through the evolutionary process and it is part of every human being. It means that it is very easy to communicate an emotion: we immediately recognize the meaning of another human's emotion. On the contrary, communicating a theorem is not easy at all, and often requires special skills.

Emotions have the important function of speeding up communication of crucial information among members of the same species.

If emotion is, ultimately, a reaction to a situation in the environment, it can be assumed to be a "measure" relative to that situation, and what is communicated is precisely that measure. But a measure of what? Buck thinks that emotions always originate from motives that must be satisfied: the emotion is a measure of how far they have been satisfied. For example, fear is a measure of safety.

A more appropriate way of referring to adaptation is "homeostasis", which is the process of searching for a balance. If something changes in the environment, all organisms that depend on that environment will somehow react to recreate the equilibrium they need to survive. This process of continuous search for equilibrium is called "homeostasis". Buck argues that the ultimate function of emotions is homeostasis.

Emotion as Heterostasis

A different view is held by the US computer scientist Harry Klopf: organisms are not hiding in the environment, trying to minimize action and change; they actively seek stimulation. If homeostasis is the seeking of a steady-state condition, "heterostasis" is the seeking of maximum stimulation. According to Klopf, all parts of the brain are independently seeking positive stimulation (or "pleasure") and avoiding negative stimulation (or "pain"). Klopf also thinks that cognition and emotion coexist and complement each other, but their relative roles are significantly different: emotion provides the sense of what the organism needs, while cognition provides the means for achieving those needs.

Emotion as Cognition

The common theme underlying all of these studies is that emotions are not as irrational as they seem to be; quite the opposite, actually. William

James explained emotions as bodily upsets, but the variety of emotional responses makes it difficult to devise a common theory of emotions (and why only some bodily upsets result in emotions). If emotions have a "cognitive life", on the other hand, that would explain their complexity.

The US psychologist George Mandler views emotion as a cognitive summary of sorts. Let's assume that, of all the information available in the environment, the mind is mainly interested in environmental regularities. Then most of its processing can be reduced to: there is a goal (e.g.: "eat"), there is a need (e.g.: "food") and there is a situation (e.g.: "a plantation of bananas"). Based on known regularities of the environment, the mind can determine what it needs to do in order to achieve its goal in the current situation. The emotion (e.g.: "to desire bananas") simplifies this process. The function of emotions is to provide the individual with the most general view of the world that is consistent with current needs, goals and situations.

The US psychologist Richard Lazarus agrees that the ultimate goal of our emotions must be to help the organism survive in the environment. Emotions arise from the relationship between the individual and its environment, or, better, the regularities of its environment. An emotion arises an appraisal of the situation and its consequences. For example, such an appraisal may lead to fear if the situation turns out to be dangerous. Emotions are genetically determined, but they can change during a lifetime: both biological and social variables may alter our given set of emotions, and this explains why emotions change through the various stages of life.

The meaning of each emotion is about the significance of the triggering event (the situation) for the well-being of the individual. Ultimately, emotions express the personal meaning of an individual's experience.

Each emotion is defined by a set of benefits and harms in the relationship between individual and environment, and that set is constructed by a process of appraisal. Appraisal is key to emotion. Each type of emotion is distinguished by a pattern of appraisal factors.

Since appraisal is the fundamental process for the occurrence of emotion, Lazarus believes that cognition is a prerequisite for emotion: a cognitive process (an appraisal) must occur before one can have an emotion.

Similarly, the Dutch psychologist Nico Frijda viewed emotions as awareness of "action tendencies", tendencies to act based on the situation.

The US psychologists Peter Salovey and John Mayer ("Emotional intelligence", 1990) introduced the term "emotional intelligence" and identified four kinds of emotional intelligence: recognizing emotions, using emotions to facilitate reasoning, understanding the meaning of one's emotions, managing one's emotions. Several scholars associated emotional intelligence with the development of empathy from the US pediatrician Martin Hoffman ("Development of Prosocial Motivation", 1982) to the US psychologist Nancy Eisenberg ("Empathy and Sympathy", 2000). In fact, the French neurologist Jean Decety and the US neurologist Philip Jackson have shown that in most cases empathy employs the same neural structures of emotional intelligence ("A Social-Neuroscience Perspective on Empathy", 2006).

Following Martin Heidegger (who thought that emotions tune us to the world) and Jean-Paul Sartre (who thought that emotions have a purpose and we are responsible for them), the US philosopher Robert Solomon argued that we are responsible for our emotions, and, in fact, we "are" our emotions (as well as our thoughts). Emotions are an essential part of our existence: without them, we would not be able to make rational decisions. It is our emotions that guide us in this world. Anger, for example, is a strategy for engaging with the world. People enjoy dramatic films and even horror films because they evoke unpleasant emotions. People even enjoy (and pay for) experiencing extreme danger (whether on rollercoasters or paragliding). We wouldn't get pleasure out of a "negative" emotion unless that emotion was not negative at all. Pleasure and pain are not opposites: they are complementary. Emotions help us conceptualize and evaluate, and therefore shape our lives. Emotions are not inside our mind but are outside, in the world, and more precisely in the social space. "Introspecting is looking in the wrong place". Emotions, therefore, help us "reason". People whose emotional life has been damaged (e.g. by a stroke) are no longer capable of making rational decisions despite the fact that the rest of their brain is functioning like before. They do not "care" for the consequences of the decision and therefore are incapable of making a rational one. Solomon also argued that spirituality is a meta-emotion that transcends the personal and relates to a larger self.

Emotion as Communication Between the Brain and the Self

Another synthesis that brings consciousness and the body into the picture has been proposed by the Spanish anthropologist José Jauregui. Jauregui, like Edward-Osborne Wilson, views sociology as a branch of biology. In his opinion, the same emotional system controls social, sexual and individual behavior. Such an emotional system originates from the neural organization of the brain: emotions are rational and predictable events. Jauregui believes that the brain is basically a computer, and emotions represent the output of that computer's processing activity. It is emotion, not reason, that directs and informs the daily actions of individuals.

Jauregui begins by separating the brain and the self: the brain is aware of what is going on in the digestive system of the body, but will inform the self only when some correction/action is necessary. Normally, an individual is not aware of her digestive processes. Her brain is always informed, though. When her awareness is also required, an emotion is generated. The communication channel between the brain and the self is made of emotions. The brain can tune the importance of the message by controlling the intensity of the emotions. The more urgent the message, the stronger the emotion. Far from being an irrational process, the emotional life of an individual is mathematically calculated to achieve exactly the kind and degree of response needed for the well-being of the individual. Feelings are subjective and inaccessible, but they are also objective and precise.

When it receives a message in the form of an emotion, the self has no idea of the detailed process that was going on in the body and of the reason why that process must be corrected. The emotion makes detailed information redundant because the emotion basically contains its own correction mechanism. The brain's emotional system is a sophisticated and complex information-processing system. The brain is a computer programmed to inform the self (through emotions) of what must be done to preserve her body and her society. It is through emotions that the brain informs the self of bodily situations that are relevant for survival.

The self maintains a degree of freedom: while it cannot suppress the (emotional) messages it receives from the brain, it can disobey them. The brain may increase the intensity of the message as the self disobeys it, and a painful conflict may arise. The brain and the self are not only separate: they may even fight each other.

In conclusion, only the self can be conscious and feel, but the brain has control over both consciousness and feelings.

If we one views the brain as a computer, the hardware is made of the neural organization. There are two types of software, though: "bionatural" (knowledge about the natural world) and "biocultural" (such as a language or a religion). A program has three main components: the sensory, the mental and the emotional systems. Any sensory input can be translated automatically by the brain into a mental (idea) or emotional (feeling) message; and viceversa. Both biocultural and bionatural programs exert emotional control over the body.

Jauregui distinguishes five systems of communication: the natural system (in which the sender is a natural thing, such as a tree), the cultural system (the sender is culture, something created by humans), the somatic system (the sender is the individual's own body), the imaginary system (the sender is imagination) and the social system (the sender is another individual). The human brain is genetically equipped to receive and understand all five kinds of messages. What ultimately matters is the emotional translation of sensory inputs arriving via these communication channels.

Emotion as Body Representation

The Portuguese biologist Antonio Damasio focused on the relation between memory, emotions and consciousness. He made a distinction between emotions and "feelings". A feeling is the private experience of an emotion, that cannot be observed by anybody else. An emotion is the brain process that we perceive as a feeling. An emotion can be observed by others because it yields visible effects (whether the facial expression or a movement) and because it arises from a brain process that can be observed and measured.

The difference is crucial. Emotions are fixed genetically, to a large extent: evolution has endowed us with a basic repertory of emotions that help us survive. My personality (which is mostly shaped by my interaction with the environment) may determine how i express and react to those emotions, but the emotions that occur in me are those that i share with my whole species. Emotion is a genetically-driven response to a stimulus: when that stimulus occurs (for example, a situation of danger), a region of the brain generates an emotion (fear) that is spread through the brain and the body via the nervous system and therefore causes a change in the state of both the brain and the rest of the body. This change of state is meant to somehow cope with the stimulus. Some emotions are acquired during development (e.g., through social interaction) but they too are grounded in the universal, primary emotional repertory of the species.

Therefore the relationship between the individual and the environment that has been posited by many thinkers as the cause of emotions is reduced by Damasio to the interaction between the body and the brain, which is only indirectly related to the interaction between the organism and the environment. Emotion is, indeed, about homeostatic regulation, is indeed about maintaining equilibrium, but the equilibrium is, more specifically, between external stimuli and internal representations.

Feelings, on the contrary, are "perceptions", except that they are a special kind of perceptions. Damasio argues that feelings are views of the body's internal organs. This follows from his theory of what the function of the mind is: the mind is about the body. The neural processes that I experience as "my mind" are about the representation of my body in the brain. Mental life requires the existence of a body, and not only because it has to be contained in something: mental life "is" about the body.

Feelings express this function of the mind. This also explains why we cannot control the feelings of emotions: we cannot because we cannot change the state of our body, or, better, we can control emotions to the extent that we can change the state of our body that caused that emotion.

Of course, that representation of the body is always present in the brain, but it is mostly dormant. It takes a specific stimulus to trigger it and generate an emotion, which in turn yields a feeling.

William James had already argued that feelings are a reflection of a change in the state of the body. Damasio gave James' intuition a detailed model: first an external stimulus triggers certain regions of the brain, then those regions cause an emotion, then the emotion spreads around the body and causes a change in the state of the body, and finally the "mind" perceives that change of state as a feeling.

Since feelings are percepts, they must be considered as cognitive as any other percept, as cognitive as an image or as a word.

Damasio's intuitive argument is that the emotional system is spread throughout the body: emotions react to states of all sorts of organs (a huge number of events can trigger an emotion, say, of pain) and operate in turn on all sorts of action, from facial expression to limb movement. Emotions are not only about the brain: they are also about the whole body.

An emotion is registered by the brain when a stimulus is recognized as useful for survival or for well-being or damaging for survival and well-being. This appraisal results in bodily changes, such as quickening heartbeat, tensing muscles, etc. These bodily changes also imply that a map changes in the brain, and this change is the physical implementation of the "feeling". Damasio finds an analogy between the emotional system and the immune system. The immune system produces antibodies to fight invading viruses; or, better, the invading virus selects the appropriate antibody. An emotional response is basically the antibody that reacts to an invading stimulus, that is selected by that stimulus.

Somatic Markers

According to Damasio, the only thing that truly matters for an individual's emotional life is what goes on in the brain. The brain maintains a representation of what is going on in the body. A change in the environment may result in a change in the body. This is immediately reflected in the brain's representation of the body state. The brain also creates associations between body states and emotions. Finally, the brain makes decisions by using these associations, whether in conjunction or not with reasoning.

The brain evolved over millions of years for a purpose: it was advantageous to have an organ that could monitor, integrate and regulate all the other organs of the organism. The brain's original purpose was, therefore, to manage the wealth of signals that represent the state of the body (the "soma"), signals that come mainly from the inner organs and from muscles and skin. That function is still there, although the brain has evolved many other functions (in particular, for reasoning). Damasio identified a region of the brain (in the right, "non-dominant" hemisphere) that could be the place where the representation of the body state is maintained. Damasio's experiments showed how, when that region is severely damaged (usually after a stroke), the person loses awareness of the left side of the body. The German neurologist Kurt Goldstein had already noticed in the 1930s that the consequence of right-hemisphere lesions is indifference.

The brain links a body change with the emotion that accompanies it. For example, the image of a tiger with the emotion of fear. By using both inputs, the brain constructs new representations that encode perceptual information and the body state that occurred soon afterwards. Eventually, the image of a tiger and the emotion of fear, as they keep occurring together, get linked in one brain event. The brain stores the association between the body state and the emotional reaction. That association is a "somatic marker".

Somatic markers are the repertory of emotions that we have acquired throughout our lives and that we use for our daily decisions. The somatic marker records emotional reactions to situations. Former emotional reactions to similar past situations is

what the brain uses to reduce the number of possible choices and rapidly select one course of action. There is an internal preference system in the brain that is inherently biased to seek pleasure and avoid pain. When a similar situation occurs again, an "automatic reaction" is triggered by the associated emotion: if the emotion is positive, like pleasure, then the reaction is to favor the situation; if the emotion is negative, like pain or fear, then the reaction is to avoid the situation. The somatic marker works as an alarm bell, either steering us away from choices that experience warns us against or steering us towards choices that experience makes us long for. When the decision is made, we do not necessarily recall the specific experiences that contributed to form the positive or negative feeling.

In philosophical terms, a somatic marker plays the role of both belief and desire. In biological terms, somatic markers help rank "qualitatively" a perception.

In other words, the brain is subject to a sort of "emotional conditioning". Once the brain has "learned" the emotion associated to a situation, that emotion will influence any future decision related to that situation. The brain areas that monitor body changes begin to respond automatically whenever a similar situation arises.

It is a popular belief that emotion must be constrained because it is irrational: too much emotion leads to "irrational" behavior. Instead, Damasio found that a number of brain-damage cases in which a reduction in emotionality was the cause for "irrational" behavior.

Somatic markers help to make "rational" decisions, and help to make them quickly. Emotion, far from being a biological oddity, is actually an integral part of cognition. Reasoning and emotions are not separate: in fact, they cooperate.

Emotion as Change Of Body State

The lessons of William James and Antonio Damasio provided a new framework for the study of emotions.

The US philosopher Jesse Prinz and the US psychologist James Laird even equated emotions to higher cognitive faculties such as vision or touch, which represent the relation between the subject and the environment.

There is evidence that specific circuits in the brain are devoted to handling emotions. These regions communicate the "emotion" to the rest of the body via the bloodstream and the nervous system. The effect is to cause a change in the state of the body. So the emotion is really an "amplifier" of a signal that came from either the body itself or from the external world (itself mediated by the senses, which are part of the body). Ultimately, the emotion looks like a loop: a change of state in the body causes an emotion that causes a change of state in the body.

The state change caused by the emotion is, somehow, a direct response to the state change that caused the emotion. The emotion is trying to maintain the "status quo" in the face of destabilizing news. The emotion is a mechanism to regulate the body, and the regulation is "homeostatic" in nature, i.e. it aims at maintaining a stable state.

That "stable" state has to do, ultimately, with survival of the organism. All emotions can be reduced to the basic emotions of "pain" and "pleasure", of negative and positive reward. Both pain and pleasure guide the organism towards the actions that maximize its chances of survival.

The brain is endowed with another mechanism for survival, the one that we call "cognition". The brain analyzes the world and makes decisions about it. Emotion and cognition work towards the same goal on parallel tracks. The advantage of emotion over cognition is that it provides a short-cut: instead of analyzing every single stimulus separately, it allows the organism to react to different stimuli with the same action. Fear is the reaction to any kind of danger, even if they are completely different events. Emotion enables similar response to different stimuli, without any need to "think" about it.

The disadvantage of emotion is that sometimes the short-cut is not perfect: it may lead us to "over-react".

Where does this "short-cut" mechanism come from? If its purpose is survival of the organism, it was probably selected by evolution. Emotion encodes a logic of survival that was developed over the course of the evolution of species.

The US psychologist Peter Lang ("The Cognitive Psychophysiology Of Emotion", 1985) believes that it all started with simple emotions related to brain circuits; that two separate "motivational" systems coexist in the brain, one ("appetitive" system) leading us towards stimuli that cause pleasure, and one ("defensive" system) steering us away from stimuli that cause pain. As the brain's circuitry grew larger and more complex, these elementary emotions of pleasure and pain, that corresponded to motivations for approach and avoidance, evolved into the varied repertory of emotions of today's humans.

It is a fact that, evolutionarily speaking, the brain components that preside over emotions are older. First brains started feeling emotions, then they started thinking.

Of Representation Systems

The British philosopher Keith Stenning not only thinks that emotion and cognition cooperate (not interfere) but even that emotions are the foundation of our mental life (not just an accident of nature or an evolutionary leftover). Emotions are a way to abstract situations. Similar emotions are used to classify situations and objects into concepts and categories. Semantically speaking, emotions are the ultimate meaning. The solution to Ludwig Wittgenstein's famous paradox (we all know what a "game" is, but there is no simple definition of what a "game" is) is simple: we know what a "game" is because we know what the emotion related to a game is. Anything that elicits the same kind of emotion is a "game". We don't need to find a definition for the word "game".

By the same token, communication is but the articulation of emotions through the development of adequate representations.

By the same token, the reason it is so easy for us to learn something so difficult as language (with all its idiosyncrasies) is that language is structured according to our emotional systems. It reflects the way our emotions work.

Stenning rediscovers an obvious truth: we are not only weird systems that build representations but also weird systems that have emotions about them. His explanation for this oddity is simple: emotions "are" the implementations of those representations in our minds.

Emotion as Memory

The closer we look, the more apparent it is that emotion is not a separate subsystem of the mind, but a pervasive feature of it. It makes evolutionary sense and it plays a crucial role in our daily actions.

Emotions are key to learning and behavior, because fear conditioning imprints emotional memories that are quite permanent. The relationship between emotion and memory goes beyond fear, but fear is the emotion that has been studied more extensively. As a matter of fact, fear seems to be a common ground for (at least) all vertebrates. The effects of fear on memory are powerful.

The British psychologist John Aggleton offered a model of how memories about fearful experiences are created in the brain by interactions among the amygdala, the thalamus and the cortex. Emotional memory (stored in the amygdala) differs from declarative memory (which is mediated by the hippocampus and the cortex). Emotional memory is relatively primitive, in the sense that it only contains simple links between cues and responses. A noise in the middle of the night is enough to create a state of anxiety, without necessarily bringing back to mind full consciousness of what the origin of that noise can be. This actually increases the efficiency (at least the speed) of the emotional response.

Emotional and declarative memories are stored and retrieved in parallel. Adults cannot recall childhood traumas because in children the hippocampus has not yet matured to the point of forming conscious memories, but the emotional memory is there.

Emotions are the brain's interpretation of reactions to changes in the world. Emotional memories involving fear can never be erased. The prefrontal cortex, the amygdala and the right cerebral cortex form a system for reasoning that gives rise to emotions and feelings. The prefrontal cortex and the amygdala process a visual stimulus by comparing it to previous experience and generate a response that is transmitted both to the body and to the back of the brain.

Therefore the brain contains a reasoning system for the emotional memory and one for the declarative memory that perform in different manners and use different circuits. In a sense, we have not one but two brains, that operate in parallel on the same input but may generate completely different output.

Reward and Punishment

The theory of emotion advanced by the British psychologist Edmund Rolls is based on the assumption that the brain was designed to deal with reward and punishment.

First, Rolls argues that the brain determines which actions to carry out based on a reward and punishment mechanism. The brain is capable of computing the "value" (in terms of reward and punishment) of a sensory input. That capability comes from millennia of natural selection. For example, when the body needs food (translation: such and such a state of energy balance is insufficient), the brain generates the emotion of hunger. That motivates the body to eat. When the body has eaten enough (i.e. a proper state of energy balance has been recreated), the brain generates the emotion of satiety.

The whole brain (not just one region) is designed around the reward and punishment mechanism. The brain plans behavior that helps obtain rewards and avoid punishments, i.e. "motivated" behavior.

Rolls deals with emotions as "states produced by reinforcing stimuli" (a purely physical, neural approach).

The special role of "reinforcing" stimuli explains why only "some" stimuli elicit emotions. The ones that are "reinforcing" need to be decoded by special regions of the brain, which generate the neural processes that we call "emotions".

Rolls argues that there are different circuits for primary reinforcers (pain, touch and the smell of food) and secondary, or learned, reinforcers (the ones that become reinforced by association with a primary reinforcer, such as the sight of favorite food). The amygdala and regions of the cortex are involved in learning reinforcers and in decoding the known reinforcers.

Emotions have several functions, including: the production of an autonomic response (i.e., a faster heartbeat); the production of an endocrine response (i.e., adrenaline); the production of motivated behavior (not only specific to the event at hand but even a direction of behavior that will last a lifetime); a re-evaluation of the stimuli themselves for future use; communication to other members of the group; the storage of "important" memories.

The representation of a "reinforcer" (i.e., of an event that causes an emotion) contains information about how to calculate reward/punishment. Not surprisingly, the representation lends itself to generalization (if one tiger causes fear, all tigers should) and association.

The emotional aspect of a perception is basically due to the activation of its representation.

Patterns of neural activity in the brain correspond to patterns of stimuli in the world, i.e. to "objects". Therefore there exists a representation of an object in the brain, and such representation is the corresponding firing of neurons. Such a mental representation obtains its meaning in two ways: from the associated reward (or punishment) "value" and from its sensory-motor correspondence in the world. First, some actions are but the means to activate such representations, because living beings are programmed to achieve reward and avoid punishment, and their search for rewards (and away from punishments) is basically a quest for "activation" of those representations. A brain will work to make those representations become active. Secondly, each "object" is associated with the kind of actions that are possible and not possible with it. The same arguments apply to language: the word "banana" has meaning both because its taste has a reward value and because it is associated to the action of peeling.

The brain is designed around the mechanisms of reward and punishment. Whatever system is in charge of "action", that system must be taking the output of reward/punishment assessment as its goal: the brain as a whole wants to maximize rewards and minimize punishments, i.e. to maximize the activation of representations relative to "rewarders" and minimize activation of the representations relative to "punishers". The specific behavior is not genetically fixed: what is fixed genetically is the goal, which is to maximize reward and minimize punishment.

There are two routes to action that can be followed in relation to reward. One route is shared by humans and other primates, so it is probably the most primitive one, and is centered on the amygdala. This route is a "shortcut" that sends stimuli directly to the motor system of the body. A second route involves "planning" the action, and therefore "thinking" about it, and centers on the language system. The second route is what allows us to make "long-term" decisions rather than only short-term decisions. It is also the one that Rolls presumes is causing consciousness. Consciousness is therefore a property of the brain processing rewards, a side effect of high-level planning.

Rolls implies that consciousness arises from the capability of thinking about one's thoughts, or higher-level thinking. He goes on to suggest that this feature could have an evolutionary value, as it allows people to correct mistakes of first-order thinking (in other words, to learn). Language is therefore necessary but not enough for consciousness: consciousness also requires the ability to think about thought, or higher-level thought.

Some sensory input becomes conscious (becomes "feeling") precisely because in this way it can be analyzed by higher-level thought. In a sense, the mechanism of higher-level thought would be useless if it could not use the feelings associated with sensory input.

Nature vs Nurture

The US psychologist Paul Ekman, following Charles Darwin himself, argued that emotions are basic, universal and hard-wired ("An Argument for Basic Emotions", 1962). All humans share the same emotions. His study of facial expressions of emotion across cultures convinced him that they are not determined by society: they are determined by biology.

The US neurologist Paul MacLean supported that theory based on the fact that we all share the regions of the brain that account for emotions.

The US psychologist Robert Plutchik identified eight primary emotions that all humans share from birth and that evolved to improve human fitness: anger, fear, sadness, disgust, surprise, anticipation, trust, and joy. In fact, all animals share these primitive "emotions": they are required forms of adaptations to the environment.

However, it seems self-evident that emotions differ from society to society. Either cultures shape emotions or viceversa: people from different regions of the world sometimes collide because they have a different repertory of values, and they have different ideas about which actions are offensive or polite.

One solution is to assume that there exists a hard core of basic emotions shared by everybody and a less rigid "fruit" of emotions that are created by culture. The British philosopher Peter Goldie called it the "avocado-pear" model and argued against it. He instead proposed a "clay model": there is a window of opportunity when emotions can be shaped, and later they become hard-wired and immutable.

The Australian philosopher Paul Griffiths proposed that emotion (as well as the rest of what we call "mind") is the result of the interaction between genetic and environmental inputs.

The US anthropologist Clifford Geertz, on the other hand, argued that, essentially, emotions are cultural artifacts. So did the New Zealand-born philosopher Rom Harre and the US anthropologist Catherine Luz ("The Anthropology of Emotions ", 1986). According to them, emotions are not universal: they are shaped by culture and socially constructed.

Affective Computing
The history of "affective computing" began in earnest with the US computer scientist Rosalind Picard (1997).

Once you have a computational model of emotional behavior, you can even imagine novel emotions that don't exist in nature; and that's what the Spanish computer scientist Lola Canamero did ("Modeling Motivations And Emotions As A Basis For Intelligent Behavior", 1997) before going on to build Nao (2010), a robot that can show its emotions.

Emotional Questions?

The amygdala is a major center for the creation of emotions. Animals whose amygdala was removed showed no emotions. It turns out, though, that the neurons of the amygdala are continuously generating what appear to be emotional states, just like the heart beats all the time. This goes against our belief that emotions are due to our reaction to external stimuli. Instead we seem to be producing emotions all the time, regardless of the external stimuli.

Are emotions just like phenotypic variations, antibodies and neural connections? Are they produced randomly all the time and then the environment (the situation) "selects" which ones get to survive?

Could emotions be yet another Darwinian system?

Emotions are useful for my survival in the world. But do "I" also have to "feel" the emotion? Couldn't the brain just send a signal to the organs without bothering "me"? Why am I aware of it?

A possibility is that being aware of an emotion means that the self can preempt the mechanic activation of a response in cases in which it would be counterproductive. Sometimes fear or hunger can lead us to actions that we may regret. If we were not aware of our emotions, we would not be able to stop the consequent actions.

Do emotions need a brain to occur? Presumably they don't need a brain as complex as ours. I feel pain in my foot. I feel anguish in my heart. There really isn't any need for an additional piece of body. If the brain is the place where emotions communicate with the "I", then that would explain why emotions also need a brain.

How did we come to build such complex emotions as, say, love? Love for a child is relatively easy to explain. But love for a woman is often a rather convoluted and turbulent affair. Most of the emotions that we feel during a day would be hard to be categorized as simple "fear" or "love" or "pain". Are they "evolutionary" consequences of primitive emotions (just like a human brain is the evolutionary consequence of primitive nervous systems), which are now part of our genetic program, or are they "social" consequences of interaction with other emotional beings: are they innate or acquired? How is a complex emotion formed from more elementary emotions?

What is the advantage of building more and more complex emotions? Could it be that more complex emotions express a better balance of reason and instinct?

For A Theory Of Emotions
I have an inner life, which is not a bodily life. Within this inner life (which it is customary to call "mind") different types of things occur. I think. I feel emotions. I dream.

There appears to be a difference between emotions and thinking. Emotions are often not desired: they occur because of external stimuli. I don't have much control over them, but they are not spontaneous: I can always relate them to an external event. Emotions have no logical construct, no flow, no time dimension. They simply happen and slowly fade away or change into other emotions: their only dimension is their intensity.

The main difference between emotions and thought is that thoughts do have a time dimension and can evolve over time. Thoughts can be controlled: I can decide if I want to think or not, and what I want to think. But they can also be spontaneous, just like emotions. Both emotions and thoughts result in behavior. Therefore, my behavior is driven by both emotions and thoughts, by both controlled and non-controlled inner behavior. Thoughts also result in emotions, albeit of a different type (like depression or anxiety).

Cognition basically mediates between emotions and thought. Emotions help organize the world in the mind, and that is what thought operates upon. Each emotion changes the mind and how deeply the emotion changes the mind depends on how intense the emotion is. That "change" is a change in cognition.

Thought can also generate a change in cognition, but we can fairly assume that even thought needs to generate an emotion before a meaningful, lasting change is performed on cognition. Basically, we can assume that nothing changes in our mind unless an emotion is created. The emotion is what causes the mind to reorganize itself.

Emotion, cognition and thought seem to repeat themselves virtually ad infinitum. Senses cause sensations, which cause cognitive events, which cause thought, which cause higher-level emotions, which cause higher-level cognitive events, which cause thought, which cause even higher-level emotions, etc. The process gets weaker and weaker as it moves higher and higher, and in most cases it actually never reaches the second level (in a significant way, at least). This process is a process similar to resonance that continues virtually forever, although it rapidly stops being meaningful, especially if new sensations start another chain of events.

Some emotions are localized and some emotions are not localized. The pain in my foot is localized, but my fear of death, my career ambitions and my desire of learning are not localized. Most emotions correspond to bodily needs, but some correspond to more abstract entities that have to do with thought itself. You need to be a thinking subject to desire to learn. Career ambitions refer to a vast complex system of values that has been built with thought. Even my fear of death is really a fear of "inner" death, not of bodily death, and therefore refers to thought.

Some emotions (the "bodily emotions") are localized and refer to the life of body parts. Some emotions ("inner" emotions) are not localized and refer to the inner life of thought. If thought is an evolution of emotions, then the latter are emotions about emotions.

Emotions play the key role of being preconditions to cognition and therefore to thought.

If thought and emotion are different processes, what is their evolutionary relationship? Have they always been different and separate processes, or is thought simply an evolution of emotions that happened when language enabled us to control emotion and to develop something equivalent to emotion but more subtle?

Note that the free will of the self is almost the opposite of emotions: emotions are beyond "our" control.

The machinery of "mind", or "cognition" (memory, learning, reasoning, language), is at the service of our primary inner life: thoughts and emotions (and even dreams). The machinery of "mind" is really a mediator between our primary inner life and our bodily life. I can remember an event, and then feel an emotion or think about that event. Viceversa, I may be thinking of something and recall an event. My inner life needs a physical support to be stored and retrieved. My current inner life needs a physical support to communicate with my previous inner life. The time dimension of thinking is implemented in the physical support. That physical support is the brain.

Mind As An Evolution Of Emotions

If emotions are the basic constituent of consciousness and they have "evolved" over the millennia, a possible and plausible explanation of where our mind comes from goes like this.

The earliest unicellular organisms were capable of irritability and excitability. That is the basic survival tool: look for what is good and avoid what is bad. The basic sensors of those organisms may have evolved into more sophisticated sensors, capable of more than just binary "good/bad" discrimination: a range of "emotions" was born. If consciousness (to some degree) is ubiquitous in nature, then one can assume that those "emotions" were associated with feelings, even if they were very limited.

Emotions existed from the very beginnings of life, and then they evolved with life. They became more and more complex as organisms became more and more complex.

Emotion detects and identifies meaning in the outside world and directs attention to that meaning. That represented a great evolutionary advantage.

Early hominids had feelings, although their feelings, while much more sophisticated than the ones of bacteria, were still rather basic, probably limited to fear, pain, pleasure, etc. In mammals and birds emotions were related to sounds (eg, fear to screaming). Early hominids had a way to express through sounds their emotions of fear and pain and pleasure and so forth.

Emotions were a skill that helped in natural selection. Minds were always busy thinking in very basic terms about survival, about how to avoid danger and how to create opportunities for food.

What set hominids apart from other mammals was the ability to manufacture tools. We can walk and we can use our hands in ways that no other animal can. The use of tools (weapons, clothes, houses, fire) relieved us from a lot of the daily processing that animals use their minds for. Our minds could afford to be "lazy". Instead of constantly monitoring the environment for preys and predators, our minds could afford to "relax". Out of that laziness modern consciousness was born. As mind had fewer and fewer practical chores, it could afford to do its own "gymnastics", rehearsing emotions and constructing more and more complex ones. As more complex emotions helped cope with life, individuals who could generate and deal with them were rewarded by natural selection. Emotions followed a Darwinian evolution of their own. That process is still occurring today.

Most animals cannot afford to spend much time philosophizing: their minds are constantly working to help them survive in their environment. Since tools were doing most of the job for us, our minds could afford the luxury of philosophizing, which is really mental gymnastics (to keep the mind in good shape).

In turn, this led to more and more efficient tools, to more and more mental gymnastics. As emotions grew more complex, sounds to express them grew more complex. It is not true that other animals cannot produce complex sounds. They cannot produce "our" set of complex sounds, but they could potentially develop sound systems based on their sounds. They don't need sound systems because they don't produce complex emotions. They have the sounds that express the emotions they feel. Human language developed to express more and more complex emotions. The quantity and quality of sounds kept increasing. Language trailed consciousness.

This process continues today, and will continue for as long as better tools allow more time for our minds to think. The software engineer who is the daughter of a miner is "more" conscious than her father. And his father was more conscious than his ancestor who was a medieval slave.

Consciousness is a product of having nothing better to do with our brain.

Further Reading
Aggleton, John: THE AMYGDALA (Wiley-Liss, 1992)

Birbaumer, Niles, Ohman, Arne & Lang, Peter: THE STRUCTURE OF EMOTION (Hogrefe & Huber, 1993)
Birdwhistell, Ray: Kinesics and Context (University of Pennsylvania Press, 1970)
Buck, Ross: THE COMMUNICATION OF EMOTION (Guilford Press, 1984)
Damasio, Antonio: THE FEELING OF WHAT HAPPENS (Harcourt Brace, 1999)
Damasio, Antonio: DESCARTES' ERROR (G.P. Putnam's Sons, 1995)
Damasio, Antonio: LOOKING FOR SPINOZA (Harcourt, 2003)
Darwin, Charles: The Expression of Emotions in Man and Animals (1872)
DeSousa, Ronald: THE RATIONALITY OF EMOTION (MIT Press, 1987)
Ekman, Paul: THE NATURE OF EMOTION (Oxford Univ Press, 1994)
Elster, Jon: ALCHEMIES OF THE MIND (Cambridge Univ Press, 2000)
Frijda, Nico: THE EMOTIONS (1986)
Frijda, Nico: THE LAWS OF EMOTION (Lawrence Erlbaum, 2006)
Fromm, Rich: THE FORGOTTEN LANGUAGE (1951)
Geertz, Clifford: THE INTERPRETATION OF CULTURES (1973)
Gladwell, Malcolm: BLINK (Brown and Co, 2005)
Goldie, Peter: THE EMOTIONS (Oxford Univ Press, 2000)
Goldstein Kurt: THE ORGANISM: A HOLISTIC APPROACH TO BIOLOGY (USA Book, 1934)
Goleman, Daniel: EMOTIONAL INTELLIGENCE (Bantam, 1995)
Greenspan, Patricia: EMOTIONS AND REASONS (Routledge, 1988)
Griffiths, Paul: WHAT EMOTIONS REALLY ARE (Univ. of Chicago Press, 1997)
Harre, Rom: THE SOCIAL CONSTRUCTION THEORY OF EMOTIONS (Blackwell, 1986)
Hobson, Allan: THE CHEMISTRY OF CONSCIOUS STATES (Little & Brown, 1994)
Hoffman, Martin: EMPATHY AND MORAL DEVELOPMENT (Cambridge University Press, 2000)
James, William: PRINCIPLES OF PSYCHOLOGY (1890)
Jauregui Jose: EL ORDENADOR CEREBRAL/ THE EMOTIONAL COMPUTER (1990)
Kleitman, Nathaniel: SLEEP AND WAKEFULNESS (1939)
Klopf, Harry: THE HEDONISTIC NEURON (Hemisphere, 1982)
Laird, James: FEELINGS (Oxford University Press, 2007)

Lane, Ricahrd & Nadel Lynn: COGNITIVE NEUROSCIENCE OF EMOTION (Oxford Univ Press, 2000)
Lazarus, Richard: EMOTION AND ADAPTATION (Oxford Univ Press, 1991)
Lazarus, Richard & Lazarus Bernice: PASSION AND REASON (Oxford Univ Press, 1994)
LeDoux, Joseph: THE EMOTIONAL BRAIN (Simon & Schuster, 1996)
Lutz, Catherine: UNNATURAL EMOTIONS (Univ of Chicago Press, 1988)
Mandler, George: MIND AND BODY (Norton, 1984)
Noe, Alva: ACTION IN PERCEPTION (MIT Press, 2005)
Oatley Keith & Jenkins Jennifer: UNDERSTANDING EMOTIONS (Blackwell, 1996)
Ortony, Andrew, Clore Gerald & Collins Allan: THE COGNITIVE STRUCTURE OF EMOTIONS (Cambridge Univ Press, 1988)
Panksepp, Jaak: AFFECTIVE NEUROSCIENCE (Oxford Univ Press, 1988)
Picard, Rosalind: AFFECTIVE COMPUTING (MIT Press, 1997)
Plutchik, Robert: EMOTION: THEORY, RESEARCH, AND EXPERIENCE (1980)
Prinz, Jesse: GUT REACTIONS (Oxford Univ Press, 2004)
Robinson, Jenefer: DEEPER THAN REASON (Oxford Univ Press, 2005)
Rolls, Edmund: THE BRAIN AND EMOTION (Oxford Univ Press, 1999)
Salovey, Peter & Mayer, John: EMOTIONAL INTELLIGENCE (Dude, 2004)
Scherer, Klaus: EXPERIENCING EMOTION (Cambridge Univ Press, 1986)
Scherer, Klaus: A BLUEPRINT FOR AFFECTIVE COMPUTING (Oxford Univ Press, 2010)
Solomon, Robert: TRUE TO OUR FEELINGS (Oxford Univ Press, 2007)
Stenning, Keith: SEEING REASON (Oxford Univ Press, 2002)
Stevens, Anthony : PRIVATE MYTHS (Harvard Univ Press, 1995)
Tomkins, Silvan: AFFECT, IMAGERY, CONSCIOUSNESS (1962)
Wundt, Wilhelm: OUTLINE OF PSYCHOLOGY (Entgelmann, 1896)
Zajonc, Robert: FEELING AND THINKING (1980)

The History of Consciousness

The Origin of Consciousness

Evolution has been at work for million of years through processes of trial and error that are widespread at all levels of organization, from the cells that form a body to societies of living beings. These processes of trial and error are embedded in every living organism. They are unconscious: they happen whether we want it or not. They are programs that run all the time. For example, our heartbeat and our breathing adapt to what we are doing. When we extend a hand to grab an object, that movement is controlled by a continuous process of recalibration. The life of plants and most animals might be driven purely by these kinds of processes. On the other hand, some living beings, and notably humans, have evolved the ability to be conscious of mental processes that are not (or, at least, appear not to be) purely made of trial and error. We can model a situation, think about it, and make a decision.

How does consciousness arise (in an individual) and how did it arise (in evolution)? It is a widespread belief that we, as individuals, are not born conscious, and life, as a natural phenomenon, was not originally conscious. If these beliefs are correct, when and how does and did consciousness arise?

One problem is to understand how consciousness is generated by brain processes. This is the "ontogenetic" problem of how consciousness "grows" during the lifetime of an individual. Another problem is to figure out what has consciousness and what does not have it. This is the "phylogenetic" problem of how it was created in the first place: did it evolve from non-conscious matter over million of years or was it born abruptly in one species (whether by divine intervention or because of the advent of new brain structures)?

How and when and why did consciousness develop? Opinions vary.

A Linguistic Origin

Several scientists believe that consciousness somehow owes its existence to the fact that humans evolved in a highly connected group, i.e. that it is related to the need to communicate with or differentiate from peers, i.e. it is closely related to language.

The Austrian philosopher Karl Popper thought that, phylogenetically speaking, consciousness emerged with the faculty of language, and, ontogenetically speaking, it emerges during growth with the faculty of language.

The US biologist George Herbert Mead believed that consciousness is a product of socialization among biological organisms. Language provides the medium for its emergence. The mind is socially constructed, society constitutes an individual as much as the individual constitutes society. According to Mead, the mind emerges through a process of internalization of the social process of communication, for example by reflecting to oneself the reaction of other individuals to one's gestures. The "minded" organism is capable of being an object of communication to itself. Gestures, which signal the existence of a symbol (and a meaning) that is being communicated (i.e., recalled in the other individual), constitute the building blocks of language. "A symbol is the stimulus whose response is given in advance". Meaning is defined by the relation between the gesture and the subsequent behavior of an organism as indicated to another organism by that gesture. The mechanism of meaning is therefore present in the social act before the consciousness of it emerges. Mead thinks that consciousness is not in the brain, but in the world. It refers to both the organism and the environment, and cannot be located simply in either. What is in the brain is the process by which the self gains and loses consciousness (analogous to pulling down and raising a window shade).

The US computer scientist Michael Arbib argued that first language developed, as a tool to communicate with other members of the group in order to coordinate group action; then communication evolved beyond the individual-to-individual sphere into the self sphere.

The British psychologist Nicholas Humphrey agrees that the function of consciousness is that of social interaction with other "consciousnesses". Consciousness gives every human a privileged picture of her own self as a model for what it is like to be another human. Consciousness provides humans with an explanatory model of their own behavior, and this skill is useful for survival: in a sense, the best psychologists are the best survivors. Humphrey speculates that, by exploring their own selves, humans gained the ability to understand other humans; and, by understanding their own minds, they understood the minds of the individuals they shared their life with.

The US anthropologist Terrence Deacon takes a "semiotic" approach to consciousness. He distinguishes three types of consciousness, based on the three types of signs: iconic, indexical and symbolic. The first two types of reference are supported by all nervous systems, therefore they may well be ubiquitous among animals. But symbolic reference is different because, in his view, it involves other individuals, it is a shared reference, it requires the capability to communicate with others. It is, therefore, exclusive to linguistic beings, i.e. to humans. Such symbolic reference includes the self: the self is a symbolic self. The symbolic self is not reducible to the iconic and indexical references. The self is not bounded within a body, it is one of those "shared" references.

A Practical Origin
Others see consciousness as useful to find solutions to practical problems. The Australian philosopher David Malet Armstrong, for example, argued that the biological function of consciousness is to "sophisticate" the mental processes so that they yield more interesting action.

Alas, today consciousness hardly contributes to survival. We often get depressed because we are conscious of what happens to us. We get depressed just thinking of future things, such as death. Consciousness often results in less determination and perseverance. Consciousness cannot be the ultimate product of Darwinian evolution towards more and more sophisticated survival systems, because it actually weakens our survival system.

Consciousness' apparent uselessness for survival may be more easily explained if we tipped our reference frame. It is generally assumed that humans' ancestors had no consciousness and consciousness slowly developed over evolutionary time. Maybe it goes the other way around: consciousness has always existed, and during evolution most species have lost part of it. Being too self-aware does hurt our chances of surviving and reproducing. Maybe evolution is indirectly improving species by reducing their self-awareness.

The Bicameral Mind
The studies conducted by the US psychologist Julian Jaynes (and, before him, by the German classicist Bruno Snell) gave credibility to the idea that consciousness may be a recent acquisition of our mental life, or at least that consciousness was not always what it is today, that it was and still is evolving.

By reviewing historical, archeological and biological documents from ancient civilizations, he concluded that until about 3000 years ago human beings were still devoid of consciousness. They still relied, like all other primates, on learned reactions. The people of even the most developed civilizations before 1000 B.C. (ancient Assyria, Babylonia, Mesopotamia, Egypt) were not "truly" conscious. Ancient books such as the Iliad and the Bible were composed by non-conscious minds that explains why they could not distinguish between real and imagined events. The characters of those books act unconsciously in making their decisions and always rely on "voices". They tend to speak in hexameter rhythms, which are characteristic of the automatic processing of the right-hemisphere brain.

Schizophrenics often tend to speak in the same rhythm. These stories are all action and no introspection.

Ancient people, because non-conscious, did not feel responsible for their actions. They had no concept of good and evil. They had no conscious memories. They had no interest in history (past). They had no interest in progress (future). They had no sense of themselves.

Human beings did already employ language to communicate with other human beings, and to cooperate and to build societies and civilizations, but, in each individual's head, that language did not serve as conscious thought: it served as communication between the two hemispheres of the brain. Human beings were guided not by conscious reasoning, but by "hallucinations". Hallucinations would form in the right hemisphere of the brain and would be communicated to the left hemisphere of the brain, which would then receive them as commands. This is what Jaynes refers to as the "bicameral mind". Human beings were led by these voices in making their important decisions. "God" is one manifestation of the bicameral mind. God is the main voice that would drive individual and social behavior. With the emergence of oral languages, the hallucinating voices for performing fundamental actions became standardized and consequently societies became increasingly organized.

A conscious mind appears in the Odyssey and the most recent part of the Bible, about 3000 years ago. Those writings gradually shifted from non-conscious actions to conscious decisions. In the Odyssey characters are aware of the moral and physical consequences of their actions. In the West, moral issues started spreading in written languages around the sixth century B.C. Chinese literature moved from the bicameral mind to the conscious mind about 500 B.C. with the writings of Confucius. Indian literature shifted to consciousness around 400 B.C. with the Upanisad.

At that time, the bicameral mind began breaking down under the pressure caused by the complexity of the environment (mainly, society). The hallucinated voices became confused, contradictory, and ultimately counterproductive. They no longer provided automatic guidance for survival. At the same time, the development of writing, and the permanent recording of procedures, in 2,000 B.C., progressively reduced the need for guidance from the hallucinated voices and replaced them with a much more effective means of organization. Consciousness was therefore invented by human beings through a process that entailed the loss of belief in gods and natural selection itself, which started rewarding conscious individuals over non-conscious ones.

Jaynes thinks that, today, governments and religions, and psychological phenomena such as hypnosis and schizophrenia, and artistic practices such as poetry and music, are vestiges of that earlier stage of human consciousness, when action was guided by the bicameral mind, because these are all manifestations of an instinctive tendency towards seeking directions, or, in general, automatic guidance, from others.

Today, these two minds still coexist: the non-conscious bicameral mind that seeks guidance from "authorities" for important decisions in complex situations (such as those related to society); and the conscious mind that creates its own decisions in more local and manageable conditions.

Jaynes' concept of consciousness was revolutionary. First of all, intelligence (or, more appropriately, cognitive faculties) and consciousness are not the same thing and they are only vaguely related. Consciousness is not necessary for concepts, learning, reason or even some elementary forms of thinking. Non-conscious beings can develop sophisticated civilizations.

Secondly, awareness of an action tends to follow, not precede, the action. Awareness of an action bears little or no influence on the outcome. Before one utters a sentence, one is not conscious of being about to utter those specific words.

Thirdly, consciousness is an operation rather than a thing. Consciousness requires metaphors to express one thing in terms of another. Consciousness is being able to construct one's narrative in terms of metaphors. Consciousness requires analogy to transform things of the real world into meanings in a metaphorical space. The mental space is created through metaphors and analogies.

Metaphors and analogies map the functions of the right hemisphere into the left hemisphere and make the bicameral mind obsolete. Metaphors of "me" and analogies of "i" enabled a greater understanding of the world and of other individuals. In turn, consciousness expanded by creating more and more metaphors and analogies. Ultimately, consciousness is a metaphor-generated model of the world.

Jaynes thinks that consciousness could not have been invented if language had not evolved to the point of facilitating metaphorical thinking. And, while oral languages developed around 70,000 B.C. and written languages began about 3000 B.C., metaphorical structures did not appear until about 1,000 B.C. Early writings in hieroglyphic and cuneiform forms reflect a non-metaphoric and non-conscious attitude.

A Synthesis: For a Darwinian Theory of Consciousness

A recurring theme (Michael Gazzaniga, Daniel Dennett, William Calvin, etc) is that the brain works as a Darwinian system, in which "mutations" of thought are continuously created and then "selected" by the environment (the same way that mutations of species are continuously created and then selected by the environment).

If we assume that the same general law of evolution is responsible for all living phenomena, from the creation of species to the immune system, and we admit that mind is one of them, then a possible scenario emerges.

Thoughts are continuously and randomly generated, just like the immune system generates antibodies all the time without really knowing which ones will be useful. Thoughts survive for a while, giving rise to "minds" that compete for control of the brain. At each time, one mind prevails because it can better cope with the situation.

Which mind prevails has an influence on which thoughts will be generated in the future. In practice, a mind is the mental equivalent of a phylogenetic thread (of a branch of the tree of life).

We are conscious, by definition, only of the mind that is prevailing.

In ancient times, the minds chaotically generated by the right hemisphere were simply shouted to the left hemisphere, which would act as the mediator with the environment: it would translate hallucinations into actions, and the result of actions into emotions, and emotions would either reinforce or weaken the mind in control. Emotions would select the mind.

This is more evident in children, who explore many unrelated thoughts in a few minutes: their behavior exhibits whatever the various minds produce. Later, the adult is better adjusted to select "minds" and does not need to try them all out. The adult has been "biased" by natural selection to recognize the "best" minds.

The 40 Hz radiation may simply be a way of scanning all available thoughts and of reporting emotions back to all minds (in other words, of reading the outputs of the minds, in the form of thoughts, and of feeding them new inputs, in the form of emotions).

The Prehistory of Brain

In the 1940s the British anthropologist Kenneth Oakley speculated that there may be three levels of consciousness, corresponding to the three evolutionary layers of the brain: awareness, controlled by the older part of the brain and related only to conditioning; consciousness, controlled by the cortex and the hippocampus, and related to the internal representation of the world; and self-awareness, due to the most recent layer of the brain and related to the internal representation of one's internal representation.

John Eccles speculated that consciousness arose with the advent of the mammalian neocortex, about 200 million years ago, the biologist Lynn

The US paleo-neurologist Harry Jerison looked at the fossil record for clues on the selection pressures that led to increases in the size of the primate brain.

Mammals evolved about 200 million years ago as the "nocturnal" reptiles. Unlike reptiles (such as dinosaurs), whose cognitive life was based on stimulus-response, mammals were capable of using sound to create a cognitive map of their environment. When the big reptiles disappeared 70 million years ago, vision too became a major source of information for the mammal brain, which evolved accordingly. In particular, the size of the brain increased dramatically. The brain of mammals was flooded with sensory inputs, and had to develop the ability to recognize an object that could be defined by many (virtually infinitely many) different sets of inputs. The solution was to develop a way to represent the perceptual world and use that representation to recognize objects. Thus the mammalian brain developed the ability to process stimuli by means of a "conscious" perceptual world, as opposed to the reflexes of the reptilian brain.

The function of consciousness was therefore to create the perception of the object, regardless of what sets of inputs originated the recognition.

The reptilian brain was simply "reacting" to stimuli, without any awareness of what those stimuli "meant". The mammalian brain was capable of transforming the stimuli into an "object" existing in time and space, and then "act" accordingly.

Jerison speculates that the human brain is, first and foremost, a marvel of integration. The brain is flooded with sensory data. If the brain had to analyze them one by one in isolation, it would be virtually impossible to cope with the number of sensory data. Jerison believes that the nervous system constructs a model of the world, and then uses that model to "understand" sensory data. The key to constructing the model of the world is to integrate all the sensory data themselves. As the model gets refined, it also gets easier to recognize sensory data for what they are. A sensory datum is not recognized in isolation, but it is recognized as part of a scene. That scene, in turn, represents the integration of all the data that have been perceived.

The implication is that we are conscious of something that is not necessarily the real world, but is simply the world that we created. The "world" that we perceive is nothing more than the model that we have created. That model is not necessarily the world as it is: it is a plausible model of the world, given what we have learned so far about it.

The Prehistory Of Mind

The British archeologist Steven Mithen found evidence in ancient history that "cognitive fluidity" caused the modern mind to arise.

First came social intelligence, the ability to deal with other humans; then came natural-history intelligence, the ability to deal with the environment, and tool-using intelligence; last, language. Once the ability to fully connect all these faculties developed, the modern mind was born. Crucial for the development of the human mind was language. In particular, metaphor and analogy are the fundamental features that allowed the human mind to develop as it is.

Homo Sapiens Sapiens appeared 100,000 years ago and initially behaved like Neanderthals, showing little intelligence. Two momentous transformations in human behavior occurred with art and technology (60,000 years ago) and with farming (10,000 years ago).

In order to explain these breakthroughs, Mithen resorts to Jerry Fodor's modular model of the mind. Initially, human minds were dominated by a general-purpose form of intelligence. Then a module appeared that was specialized for socializing. The social-intelligence module was shared with other primates so it must have predated humans. Then other modules, each specific to one domain, were born around the main general-purpose module. The modules evolved separately. Eventually, Mithen admits four types of intelligence (four modules in the mind): social, technical (tool-making, house building), natural-history (e.g., animal behavior) and linguistic. These modules were not connected, these "intelligences" were not communicating.

Mithen can thus explain why there is no archeological evidence of social life when (judging from brain size) social intelligence must have been already quite developed: a cognitive barrier between social and technical intelligence made it impossible for humans to conceive of tools for social interaction. Originally, humans were hunters and gatherers (the transition to farming occurred in the Middle East only about 10,000 years ago). The hunter-gatherers of our pre-history were experts in many domains, but those different kinds of expertise did not mix, precisely because the minds of those humans could not mix different types of intelligence.

"Cognitive fluidity" (mixing different kinds of intelligence) changed that and caused the cultural explosion of art, technology, religion. Suddenly, humans acquired minds in which modules had been connected. For example, tools started being used to transform nature. Religion was a by-product of mixing these intelligences, because mixing intelligences one can produce supernatural beings.

Farming was also a product of cognitive fluidity and in turn caused a redefining of intelligences (emergence of new intelligences, disappearance of old ones).

The factor that contributed or caused cognitive fluidity may have been the dawning of consciousness. Self-awareness may have integrated intelligences that for thousands of years had been kept separate.

Mithen's evolutionary theory mirrors in many ways the theory of child development advanced by British psychologist Annette Karmiloff-Smith.

The Origins of Civilization

Sometime in the neolithic past, humans discovered agriculture. At about the same time they started creating cities. Beliefs coalesced around religions and political structures arose. One way to look at this story is that the transition from hunter-gatherer to farmer caused a new way of thinking that yielded religion and cities.

Another way to look at this story is that a new way of thinking (due to a physical change in the structure of the human brain) led to religion and philosophy, and this new way of thinking manifested itself in agriculture and cities.

A very small change in the genome can cause a very big difference in the brain (after all humans share 98.5% of their genes with chimps, and the genome of Homo Sapiens is virtually identical to the genome of the Neanderthals) and a very small change in the brain can cause a very big difference in behavior.

The traditional narrative is that humans discovered agriculture and created cities, and then religion/politics evolved because agriculture/cities fostered a new way of thinking, and this new way of thinking was more rational than the previous one; but one can also view it the other way around.

For whatever reason the bodies of Homo Sapiens change over the centuries. The most visible feature is the height: we are way taller than neolithic people. There might also take place more subtle changes in the brain. When the brain changes in some individuals, the other individuals call it "madness". However, when the change is caused by diet, pathogens or some unknown biological law, it can spread just like any epidemic. As more and more individuals acquire the new brain, eventually they come to rule: their brain is now the "modern" brain, and their thinking is superior to the old thinking of the "traditional" brain.

The new brain causes a new way of thinking, initially viewed as madness but later simply accepted as "modern". The modern way of thinking causes new behavior. Eventually only the children born with this new brain are accepted and reproduce. The others die away.

According to this alternative narrative, at some point the "modern" brain of the neolithic individual started forming symbolic systems that we now call "religion" and "politics". That new way of thinking caused a change in behavior, from hunting/gathering to agriculture and cities, not because it is more rational and efficient, but simply because their brains started thinking that way. Needless to say, a brain tries to prove to itself that it is the best brain ever. (Whatever you think the brain is, it's your brain talking about itself). The new brains convinced themselves that the transition to agriculture was the right thing to do and that such a transition constitutes "progress", that it was rational discovery when in fact it was irrational self-delusion.

Because the archeological record does not show which one happened first (agriculture/cities or religion/philosophy), historians assumed that agriculture happened first; but it may well be the other way around: first our brains started (accidentally) believing in deities of fertility, rain dances and river spirits; and then we started farming and creating cities. That is idea that the French archaeologist Jacques Cauvin proposed.

In that case we may have been doing this for thousands of years. We think it has been constant rational progress, i.e. better adaptation to the environment, but in fact changes in behavior have been driven by random changes in the brain; which means that sometimes we adapt better and sometimes we don't. But every time our brains convince themselves that we adapted better.

Co-Evolution Of Language And Consciousness

The British psychologist Euan MacPhail is another scholar who believes that consciousness comes from language, and therefore it is unique to humans.

"Association formation" is ubiquitous in vertebrates, and it forms the basis for every form of learning. But humans differ from animals in that humans are capable of language, humans possess an innate ability for acquiring language.

MacPhail relates this fact to memory structures, and he does so by unifying two findings about memory.

On one hand, he thinks that humans are endowed with two parallel learning systems: a conscious (explicit) and an unconscious (implicit) system, corresponding to two memory systems, one unconscious and one conscious. The unconscious learning system is the human analogue of an animal's associative learning system. While they are both present at all times, we cannot consciously recall episodes stored in unconscious memory, whereas we can consciously recall episodes stored in conscious memory. Conscious memory develops with language, and that explains why we cannot recall episodes of our early life.

On the other hand, conscious memory is an "autobiographical" memory in the sense that it develops as the concept of "self" develops. I can feel pain only after i have developed a concept of "i", only after i have come to realize that i am myself. What feels the pain is the network of neurons that constitutes the self.

By merging the two aspects of conscious memory, MacPhail reaches the conclusion that other animals only have the implicit (unconscious) kind of memory and learning, whereas humans developed also the explicit (conscious) kind, and the latter requires the development of the self.

The origin of consciousness is therefore predicated on the origin of the self. The self, in turn, is a by-product of "aboutness", which is a requirement and a by-product of language.

The association between a subject and a predicate in language is structurally different from the associations that animals are capable of. Animals can learn associations between stimuli, but cannot infer subject-predicate associations, and that is the prerequisite to acquiring a language. Language allows humans to think in terms of "representations", of "aboutness", of the philosophical "intentionality" (from "intendo", i.e. being able to refer to something else). Animals, who are not endowed with language, cannot grasp this "aboutness". The "aboutness" relationship is the fundamental grammatical requirement for language. It is the ability to deal with "aboutness" that enables the formation of a concept of self. It is the concept of self that enables consciousness. The ability to create relationships of "aboutness" mature in children and leads to a conception of the "non-self", which in turn is reflected in a conception of the "self". At this point conscious memory starts developing, and conscious recall is possible, and conscious life begins. Consciousness is the consequence of the evolution of "aboutness".

Inasmuch as "aboutness" is the key to consciousness, Brentano was therefore correct: intentionality is the fundamental property of mind, that distinguishes it from matter.

MacPhail believes that language, the self and consciousness develop together in the infant, and this development somehow recapitulates the evolution of language in our species: we started to think when we acquired the ability to discriminate self and non-self, and we acquired that ability when we acquired the ability to learn languages.

What remains to be explained is what causes infants to diverge from other animals. If, as toddlers, we are no more conscious than puppies, what happens to toddlers that does not happen to cubs, so that after a few years a toddler is conscious and a cub will never be? Ultimately, MacPhail postulates that the answer lies in our ability to learn languages, i.e. that something unique in the human genome sets in motion a process to learn languages that is unique to humans.

Mimesis

The US linguist Merlin Donald argued that the modern mind of symbolic thought arose from a non-symbolic form of intelligence through gradual absorption of new representational systems. The human mind developed in four stages (which, incidentally, roughly correspond to stages of cognitive growth in modern humans).

Early hominids were limited to episodic representations of knowledge, which was useful for remembering repeating episodes (the "episodic" mind). The episodic memory was useful to learn stimulus-response associations, but it could not retrieve memories independent of environmental cues. In other words, it could not "think". These "episodic beings" (still more apes than humans) lived their lives entirely in the present.

Homo Erectus developed a "mimetic" (pre-linguistic but roughly symbolic) system of motor-based representations. At this stage the mind was capable of retrieving memories independent of environmental cues, and was capable of "re-describing" experience based on the overall knowledge. This is what the British psychologist Annette Karmiloff-Smith refers to as "representational re-description" in the stages of child development. The mind has a representation of the world and it is capable of continuously adapting it to new knowledge. The mind has "understanding" of the world.

These representations also enabled the individual to communicate intentions and desires and, on a larger scale, enabled generations to pass on cultural artifacts. At this stage, there existed a sort of collective memory (a "culture") founded on the ability to carry out collective motor-based re-constructions of earlier incidents. By "motor-based", Donald means that early humans were able to use their bodies to learn, remember and teach. Tool-making and games originate at this stage.

In the third stage, Homo Sapiens acquired language and therefore the ability to construct narratives and build myths, and myths represent integrated models of the world by which individuals could generalize and predict (the "mythic" mind). This stage requires new anatomical (and, specifically, neuronal) additions to the human body. These humans were capable of telling stories, a quantum leap in communication. Thus, one of language's fundamental functions is to express myths. "Language is about telling stories in a group".

About 50,000 years ago humans began to store memories in the outside world instead of in their own brain (e.g., cave paintings, figurines, calendars, etc).

Finally, modern humans, helped by written language, achieved higher, symbolic representational capabilities such as logic (the "theoretic" mind).

According to the epistemological theories of the Swiss psychologist Jean Piaget and the Russian psychologist Lev Vygotsky, children follow a similar path to full-fledged thinking, from event to mimetic, from narrative to symbolic.

Donald's fundamental insight is that language and thought are tightly related: some forms of thought require language, and language reflects what forms of thought are possible. Symbols per se did not cause a major revolution in thinking: the kind of mental models that the mind could build caused the revolution. And language (or symbols) was simply a means to represent those models. The purpose of language was to allow individuals to share a common model of the world. Narrative was the natural product of language. Narrative led to unified, collective models of reality, in particular those embodied by myths.

Cultural Origins

The US psychologist Michael Tomasello believes that human civilization is so fundamentally different from the societies of other animals because human cognition, at some point in evolution, became a "collective", not only individual, process. This was originally a small difference, but over time it has made a huge difference, because each generation hands down a "culture" to the next generation. Each generation can benefit from the experiences (e.g., discoveries and inventions) of previous generations. And this causes an acceleration of cognitive evolution.

The Three Stages of Brain Evolution

The US developmental psychologist Stephen Porges characterized the evolution of consciousness as a transition from a state of being acted upon by the world to a state of acting upon the world. Consciousness originated when the brain evolved from the reptilian structure to the mammalian structure (using Paul MacLean's model for the evolution of the brain).

The brain of a reptile (which de facto means the brainstem and the hypothalamus) is not active, but simply reactive: it reacts to food, light, temperature. The reptilian brain increases or decreases metabolism based on the body's needs. Matter prevails over mind.

In a mammal, instead, the brainstem and the hypothalamus command adjustments so that body temperature and metabolism are kept stable. This phenomenon enables the brain to dedicate energies to other functions. The brain of a mammal is capable of acting: mammals explore their environment looking for what they need. Mind prevails over matter.

The same argument can be made from an energetic perspective, which is reflected in the differences between the reptilian and mammalian cardiac systems. In the highly competitive world of mammals, it is necessary for the body to increase the production of energy to deal with preys and predators (hunt or run). So it is no surprise that mammals have metabolic demands four to five times that of reptiles, which makes reptiles more prone to passive feeding strategies, whereas mammals can actively hunt and graze and adapt to changing environments.

The reptilian brain is designed to use food. The mammalian brain is designed to look for food.

The structures in mammals (i.e., facial muscles, larynx) that express emotion (facial expression, vocalization) were evolution of anatomical systems of the reptiles. The resulting organization of the brainstem in mammalians fostered brain functions of attention, motion, emotion, and communication.

The development of the cortex enabled the mammalian brain to communicate emotions. Then it was just a matter of time before language and conscious thought emerged.

The Evolution of Feelings

The British chemist Graham Cairns-Smith views consciousness as an evolution of elementary emotions.

First, a rudimentary system of feelings must have been born by accident. Then it must have proven to have evolutionary usefulness. Finally, from that rudimentary system, that was probably a very basic pain-pleasure system, more complex feelings evolved.

Initially, they may have been simple variations on the basic emotions of pain and pleasure (or a broad palette of feelings, from pleasant to unpleasant, as the subtlety of our five senses seem to imply). As they proved to be more and more useful for survival, more and more emotions may have popped up. Eventually the organism was flooded with emotions and something like a primitive "stream of consciousness" appeared. Verbal language simply put words to it. Language allowed us to express the stream of emotions in a more sophisticated way than the primitive facial language. Thought was born. With thought even more complex emotions were born. With language, thought and deep emotions, the conscious "i" was born.

Bottom line: consciousness originated from the evolution of feelings. Feelings begat consciousness, not the other way around.

Daniel Dennett thinks that the mind was created by the evolution of memes. Cairns-Smith thinks that the mind was created by the evolution of emotions. Where most thinkers see language as essential to the development of consciousness, Cairns-Smith views it as a mere tool to communicate emotions in a more complete way. Where most thinkers see emotion as a corollary to consciousness, Cairn-Smiths views it as the embryo of consciousness.

How Homo Became Sapiens

Likewise, the Swedish linguist Peter Gardenfors views language as the last (not first) stage in the process that led to today's conscious humans. He believes that first came sensations, then attention, then emotions, then memory, then thoughts (by which he really means "internal representations of the world"), then planning, then the self, then free will and finally language.

Most of these faculties are not unique to humans. Most mammals have emotions and even thoughts. Chimpanzees exhibit all of these faculties up to planning. But he thinks that humans are the only animals that are truly conscious of themselves and can speak.

The cortex is the place where a representation of the world is created. That allows the brain to use the representation of an object (or a situation) rather than the object (or the situation) itself. It allows, in other words, to be somewhat "detached" from reality: the brain can work on something that is not an object/situation present "here and now". Gardenfors believes that the large cortex of the human brain (i.e., its superior ability in representing the world) makes all the difference between human and animal behavior. Other animals have a cortex too, but it does not compare in size with the human cortex.

Gardenfors believes that first came sensations, then perceptions (the interpretation of those sensations, which are already representations but are directly related to the world) and then "detached" representations (which he also calls "imaginations", and differ from perceptions which are "cued" representations, i.e. representations about something that is not present here and now). Since all animals have sensations, Gardenfors assigns a degree of consciousness to all animals. But only mammals and birds have the cortex that allows for detached representations: they can "guess" and "plan". E.g., a cat does not need to see a mouse to understand that it is hiding in a place where it cannot be seen, and the cat can make a plan (by guessing how the mouse will behave) in order to catch it.

Gardenfors explains the difference between sensations and perceptions as a difference in the referent: sensations are about what is happening to the body, whereas perceptions are about what is happening in the world (that is causing that change in the body). A perception is, in a sense, a step back to find out what caused the sensation. "We perceive the causes".

Humans are better than any other animal at discovering the causes because they have better "simulators" in their cortex.

The next step up, the detached representations, are important because they can be used at any time, regardless of whether the object is present or not. They also provide an evolutionary advantage: the animal can play trial and error in its internal representation, without risking its life in the real world. The animal can simulate the consequences of acting before actually acting. An internal representation "allows our hypotheses to die instead of us". Animals that are capable of internal representation (which are animals with a large cortex) share some behavioral traits: they play and they dream. Reptiles do not play and do not dream.

Thoughts (his nickname for "internal representations of the world") allowed some animals to "become increasingly detached from the immediate vicinity". Instead of reacting directly to stimuli from the environment, these animals can use "reason" to understand what is going on in the environment and to decide what to do next. An animal that can only react directly to a stimulus is limited to one course of action. An animal that can build an internal representation of the world is capable of creating more than one possible course of action.

The next step up is to actually "plan" an action. Many animals plan, but in an "immediate" fashion. Humans can plan in an "anticipatory" manner. The difference is about being ready for the same situation to occur again in the future. For example, animals make tools to be used immediately, but only humans carry their tools with them, knowing that they may need them again. Other animals would simply make the same tools again when required. Another example is how we communicate: animals do communicate, but their communication is about the "here and now", whereas humans can discuss our memories from the past and dreams for the future. In a sense, another proof of this difference is the fact that only humans seem to be aware of the full meaning of death: they not only fear it, but are devastated by the mere thought of it (note that humans bury their dead, and this custom seems to be relatively recent in the evolution of humans).

Another ladder of cognitive abilities has to do with the kind of things that one's brain can represent: an internal representation of the world, which is necessary for immediate planning; "compassion" (an understanding of others' emotions); a theory of attention (understanding what others are focusing on); a theory of intention (understanding why others are doing what they are doing); a theory of others' minds (which is basically the ability to represent the internal representations of other minds); and finally self-consciousness (a representation of one's internal representation, which is required for anticipatory planning). The jump from understanding intentions and having a theory of others people's minds may be the most difficult one: children acquire a theory of other people at about the age of four; and it is still being debated whether apes ever do. Thus Gardenfors concludes that, in all likelihood, only humans are self-conscious.

The self, the last stage of human cognitive development, presupposes a "you". Gardenfors assign a key role even to deceit and cooperation. These are phenomena that presuppose an understanding of others people's minds. The level of sophistication that the human race can achieve in matters of deceit and cooperation is due to the ability to work with the chain of nested beliefs: "I know", "I know that you know", "I know that you know that i know", etc. When one can see one's mind through the eyes of a competitor or a partner, one is seeing one's own mind. One can see one's own internal representations. Thus Gardenfors believes that an understanding of others people's minds came before an understanding of one's own mind. I understand that you exist, act and have motives before i understand that i exist, act and have my own motives. First came the concept of "i and you", then came the concept of "i" (the subject, which presupposes a non-subject), and finally the concept of "it" (the object of the subject, which presupposes a subject).

Gardenfors believes that the self is an "emergent" phenomenon, a property of the whole that was not a property of any of its constituents. The "i" emerges from a network of inter-related cognitive functions.

Gardenfors' theory of cognitive steps is consistent with Daniel Dennett's classification of "kinds of minds": "Darwinian creatures", which only live in the present; "Skinnerian creatures", which are capable of learning from trial and error; "Popperian creatures", which can play an action internally in a simulated environment before they perform it in the real environment; and "Gregorian creatures", which can extend their cognitive functions outside their organism by using tools and language.

Gardenfors adds a fifth kind to Dennett's kinds of minds: "Donaldian" beings, named after Merlin Donald's third phase: Donald believes that about 50,000 years ago humans began to store memories in the outside world instead of in their own brain (e.g., cave paintings, figurines, calendars, etc). The invention of external memories (which does not imply any change in the structure of the brain) was fundamental for creating the kind of mind that we now have. Writing and science were simply further evolutions of that invention.

Ultimately, it is all about the internal representation, which in humans is "detached" enough to allow for thinking about the past and the future, and even for thinking about ourselves.

Gardenfors sees evidence that humans have better "simulators" of the environment (building better representations) in apparently unrelated facts such as the human ability to aim and to beat a rhythm. Apes cannot aim and cannot keep time.

The consequences of having good simulators are civilizations.

Thus Gardenfors concludes that language came last: not only was it unnecessary for the birth of consciousness, but consciousness is a primitive phenomenon and language is the last stage of cognitive evolution. Human language requires a kind of internal representation (the "detached" kind) that only humans have. Basically, it requires "symbols". In a sense, human language is about what is not here and not now, whereas other animals can only communicate about here and now, because their representations are not "detached" enough from external reality (they "are" about external reality).

This limitation of other animals also translates into the sounds that they can produce. Humans are the only animals that can "choose" what sound to produce. Other animals have a repertory of sounds that they produce, and cannot control them. Humans can control them. Sometimes humans use the "instinctive" repertory of sounds (e.g., a scream or laughter). But humans can also articulate speech. Animals cannot literally talk. "They have no need to talk since they have nothing to talk about". They have no detached representation. They have no need to "talk" about things that are not here now. Gardenfors believes that even self-consciousness is required to be capable of speaking, because human language is very much about the "I" and the "you".

Gardenfors agrees with Robin Dunbar that, originally, language had a social function. Humans chatted for the same reason that apes groom each other: to cement social bonds. That helped humans create groups, and groups helped individuals survive in a hostile environment. So language had an evolutionary advantage.

Noam Chomsky's theory of an innate universal grammar is unnecessary because grammar could be (yet another) emergent phenomenon that arises after speech already existed. The brain basically organizes the speech acts that it is performing. This results (not the cause) in the rules of grammar.

Language is not handled by a separate "module" in the brain, as Chomsky claims. Instead, it is a natural evolution of cognitive skills that predated it.

Protoconsciousness

The US psychologist Stuart Hameroff advanced a theory of consciousness rooted in Physics. One of the big mysteries of evolutionary Biology is the sudden explosion of species during the Cambrian period. According to fossil records, life on Earth originated about 4 billion years ago, but for about 3.5 billion years it evolved very slowly, producing mainly single-celled organisms and a few simple multicellular organisms. Then, all of a sudden, in a rather brief period of 10 million years beginning about 540 million years ago (the Cambrian period), a huge number of different forms of life emerged. Biologists have always been puzzled by this sudden diversification of life.

One possible explanation would be the emergence of a feature that greatly enhanced adaptation and mutation. Hameroff thinks that it may have been the emergence of consciousness, that consciousness not only occurred early in the evolutionary path but it even altered the course of evolution. The idea is that behavior can indirectly alter genetic information, as already argued in 1958 by the Austrian physicist Erwin Schroedinger, by enabling organisms to survive and reproduce where non-intelligent organisms would simply die.

Cells contain a structure called "cytoskeleton", which is made of a protein called "tubulin", which forms cylinders called "microtubules". According to the US biologist Lynn Margulis, microtubules and the cytoskeleton were created by symbiotic mergers more than a billion years ago. Simple organisms actually had to rely on the cytoskeleton for purposeful behavior. Having no synapses or neural networks, they relied on their cytoskeleton for sensation, locomotion and information processing. Cytoskeletal structures provided several services, including internal organization of the neuron, processing of information, communication. In summary, the cytoskeleton organizes intelligent behavior in simple organisms. Margulis thinks that consciousness was a property of even simple unicellular organisms of several billion years ago.

The cytoskeleton seems to play a particularly relevant role in differentiation. A cell's genes are activated and regulated by its cytoskeleton. Cytoskeletal cooperation among neighboring cells enabled differentiation and allowed different types of tissues to emerge. Then, higher order structures appearing with specific functions (organs) started appearing and these in turn enabled more purposeful behavior.

All of this depends on the cytoskeleton, which Hameroff thinks is the level at which consciousness is created. If that is the case, then rudimentary "conscious" events occurred the very moment the cytoskeleton became important for small organisms. Organisms began to experience feelings and make conscious choices.

The End of the Struggle and the Luxury of Consciousness

A modest proposal: i think that consciousness came with the end of danger.

The human mind (cognitive faculties plus consciousness) was just an organ of the body, useful like the others to survive in the environment. Where the hand was useful to grab things and the leg was useful for running, the mind was useful for deciding what to do in the face of danger. The mind was capable of organizing knowledge about the world and relating it to bodily needs (food, sex, shelter, etc). In a hostile and unpredictable environment, the mind was presumably busy all the time with practical chores. As humans became less and less vulnerable to natural selection, the mind became less and less "useful". Nonetheless, the mind was still collecting and organizing knowledge about the world. Once survival got easy, the mind had "spare time" to spend with its knowledge. (We can expect that domestic animals will also go down the same path of increased awareness, as they become pets and are sheltered from their ecosystem's selection).

The human mind works in two dimensions: 1. It uses whatever knowledge it has to determine behavior in the environment; 2. It uses whatever spare time it has to refine and increase its knowledge. From knowledge better and better knowledge can always be created. That is what the mind does when it is freed from the need of concerning itself with survival. The more knowledge gets created, the more efficient the mind will be the next time it has to deal with a matter-of-life-or-death situation. That is why it takes advantage of every "break" to increase its knowledge. If the mind is "inactive" (as far as struggling for survival goes), then knowledge keeps increasing exponentially, in all directions. That includes knowledge about the mind itself. The mind becomes more and more aware of its own existence.

The mind was in origin just one of the body's features, caused by one of the body's organs (the brain), just like "walking" is a body's feature caused by a body's organ (the leg). As knowledge about itself increased, the mind became more and more independent of the environment's conditions, more and more independent of the body's needs, more and more a machine to acquire and process knowledge, more and more a feature about itself.

Consciousness comes with the end of the mind's usefulness. As the mind becomes useless, while its brain processes are unstoppable, it turns into higher and higher degrees of self-awareness. Instead of using knowledge to analyze the world, recognize natural patterns, predict situations and mandate behavior, the mind uses knowledge to create more knowledge. Eventually, it also creates more and more knowledge about itself.

A Darwinian History of Consciousness

The Italian mathematician Piero Scaruffi ("A simple theory of consciousness", 2001) views consciousness as originating from the co-evolution of memes, language, tools, emotions and brains.

If one applies Darwinian thinking to the origins of consciousness, one is led to believe that today's consciousness must be a point in a continuum of consciousness that started a long time ago and underwent evolution. If we accept that the human mind is just one of the organs that evolved over millions of years, the origins of the mind must be found in 1. A primordial organ of "thinking" and 2. An evolutionary advantage of that organ that made it evolve into what it is now.

It is likely that a number of facets of our experience evolved together.

First of all, we are a tool-making species. And tools have always shaped the mind. We are not the only tool-making species, and we are not the only species whose "cognitive life" is shaped by tools. Even a spider, that has built a web, will have a "mental" life that revolves around the latter. Each new tool, whether fire or television, has shaped the mind of the humans who used it. Tools contribute to create the mind as it is because they change the environment in which the mind must operate. As tools have evolved, from the wheel to the automobile, according to a Darwinian scenario of their own, our mind has evolved with them.

Secondly, the primordial "mind" that evolved from non-conscious matter ages ago is likely to have been very simple, possibly limited to a few emotions. For example, it may have only been capable of feeling pain and pleasure. Those emotions proved to have an evolutionary advantage, and therefore they reproduced and eventually evolved into more complex emotions, such as fear and desire. And so forth: as they proved more and more useful for survival and reproduction, eventually a whole spectrum of emotions began to emerge.

Emotions had an evolutionary value, as they helped bodies (and their genes) survive, and therefore were valuable, and therefore evolved. It is unlikely that humans are the only species with emotions, but it is likely that humans are the species in which emotions evolved in the most spectacular way. The reason for this spectacular evolution may very well be that at the same time we were developing ever more sophisticated tools than any other species. Tools relieved us from many daily chores. Our emotions had been invented to help cope with those chores, but, thanks to tools, our emotions gradually became less and less crucial to survival. The fear of tigers is important to survive in a tiger-rich environment, but once we build fences around our dwelling that emotion becomes less crucial; at least, we don't need to fear tigers all the time.

Our mind was nonetheless still producing emotions, because once an organ is created that does something it will continue to do that something. We can't just turn off our immune system because this morning there are no viruses around. Just like the immune system is producing antibodies all the time, the mind is producing emotions all the time. That flow of "free" emotions eventually led to what we call "thought". Thought eventually yielded a continuous flow of emotions and a concept of self: consciousness was born. Consciousness was born because our mind had nothing to do most of the time. We became conscious because we had nothing better to do with our emotions.

At the same time, communication was also evolving. Language evolved from primitive sounds and gestures because, again, it provided an evolutionary advantage. Language shaped the mind as much as the mind shaped language. The very idea of the "self" may have originated from the ability to think in a structured manner about our experience, the ability to form narratives.

Finally, memes evolved. Ideas, slogans, religions, ideologies evolved from the early, very basic, concepts of the world. And, again, memes shaped the mind as much as the mind shaped memes.

Today's mind appears to be the result of the co-evolution of brains, tools, emotions, language, memes.

It was evolution on several parallel tracks.

Further Reading
Arbib, Michael: METAPHORICAL BRAIN 2 (Wiley, 1989)
Cairns-Smith, A. G.: EVOLVING THE MIND (Cambridge University Press, 1995)
Cauvin, Jacques: THE BIRTH OF THE GODS AND THE ORIGINS OF AGRICULTURE (Cambridge Univ Press, 2000)
Deacon, Terrence: THE SYMBOLIC SPECIES (W.W. Norton & C., 1997)
Donald, Merlin: ORIGINS OF THE MODERN MIND (Harvard Univ Press, 1991)
Gardenfors, Peter: HOW HOMO BECAME SAPIENS (Oxford Univ Press, 2003)
Hameroff, Stuart: ULTIMATE COMPUTING (Elsevier Science, 1987)
Humphrey Nicholas: CONSCIOUSNESS REGAINED (Oxford Univ Press, 1983)
Jaynes, Julian: THE ORIGIN OF CONSCIOUSNESS IN THE BREAKDOWN OF THE BICAMERAL MIND (Houghton Mifflin, 1977)
Jerison, Harry: THE EVOLUTION OF THE BRAIN AND INTELLIGENCE (1973)
MacPhail, Euan: THE EVOLUTION OF CONSCIOUSNESS (Oxford UnivPress, 1998)
Mead, George Herbert: MIND, SELF AND SOCIETY (Univ of Chicago Press, 1934)
Mithen, Steven: THE PREHISTORY OF THE MIND (Thames and Hudson, 1996)
O'Flaherty, Wendy: KARMA AND REBIRTH IN CLASSICAL INDIA (University of California Press, 1980)
Oakley, Kenneth: MAN THE TOOL-MAKER (1949)

Popper, Karl & Eccles, John: THE SELF AND ITS BRAIN (Springer, 1977)

Popper, Karl: KNOWLEDGE AND THE BODY-MIND PROBLEM (Routledge, 1994)

Schroedinger, Erwin: Mind and Matter (1958)

Snell, Bruno: THE DISCOVERY OF THE MIND (1946)

Tomasello, Michael: THE CULTURAL ORIGINS OF HUMAN COGNITION (Harvard Univ Press, 1999)

Tull, Herman: THE VEDIC ORIGINS OF KARMA (State University of New York, 1989)

Consciousness: The Factory of Illusions

Science's Last Frontier

The 20th century witnessed tremendous scientific progress in many fields. This has brought about a better understanding of the world we inhabit, of the forces that drive it, of the relationships between the human race and the rest of the universe. Scientific explanations have been provided for most of the phenomena that used to be considered divine events. We have learned how the universe was born, and how it gave rise to the galaxies and the stars and ultimately to our planet; and what life is, how it survives, reproduces and evolves; and what the structure of the brain is, and how it works.

The mystery is no longer in our surroundings: it is inside ourselves. What we still cannot explain is precisely that: "ourselves". We may have a clue to what generates reasoning, memory and learning. But we have no scientific theory for the one thing that we really know very well: our consciousness, our awareness of being us, ourselves.

No scientific theory of the universe can be said complete if it doesn't explain consciousness. We may doubt the existence of black holes, the properties of quarks and even that the Earth is round, but it is harder to doubt that we are conscious. Consciousness is actually the only thing we can be sure of: we are sure that "we" exist, and "we" doesn't mean our bodies: it means our consciousness. Everything else could be an illusion, but consciousness is what allows us to even think that everything else could be an illusion. It is the one thing that we cannot reject.

If our theory of the universe that we have does not explain consciousness, then maybe we do not have a good theory of the universe. Consciousness is a natural phenomenon. Like all natural phenomena it should be possible to find laws of nature that explain it.

Unfortunately, precisely consciousness, of all things in the universe, still eludes scientists. Physics has come a long way to explaining what matter is and how it behaves. Biology has come a long way to explain what life is and how it evolves. But no science has come even close to explaining what consciousness is, how it originates and how it works.

Neurology tells us an enormous amount about the brain, but it cannot explain how conscious experience arises from the brain's electrochemical activity.

One wonders if there is still something about the structure of matter that we are missing.

We may have figured out the meaning of matter and the structure of life, but we were more interested in the structure of matter and the meaning of life.

What Is Consciousness?
What is consciousness? What is it to be aware? The more we think, the less we can define it. How does it happen? How does something in the brain (it is in the brain, isn't it?) lead to our emotions, feelings and thoughts? And why does it happen? Why were humans (and presumably, to some extent, many other animals) endowed with consciousness, with the ability to know that they exist, that they live, that other people live, that they are part of this universe and that they will die? Why do we need to "think" at all?

Why doesn't our inner life mirror faithfully, one to one, our external life? When we experience sensations related to interactions of our body with the world, our emotional life can be said to mirror the environment. But when we think, sometimes we think things that never happen and will never happen. How can consciousness be so decoupled from the environment if brain processes are tightly coupled to it? Is consciousness a form of self-maintenance the same way that the autonomic system is a form of self-maintenance of the body regardless of what happens in the environment?

Paradoxes and weird properties of consciousness abound.

Why can't i be aware of my entire being? We only have partial introspection. We have no idea what so many organs are doing in our body.

Consciousness is limited to my head. Do I need hands and feet in order to be conscious? Is consciousness only determined by what is in the head, or is it affected also by every part of the body? Am I still the same person if they cut my legs? What if they transplant my heart?

We can only be conscious of one thing at a time. There are many things that we are not conscious of. How do we select which thing we want to be conscious of?

Why can I only feel my own consciousness and not other people's consciousness? Why can't I feel other people's feelings? Why can't anybody else feel my feelings? Conscious states are fundamentally different from anything else in nature because they are "subjective" and "opaque" (i can't feel yours). They are not equally accessible to all observers.

Consciousness is a whole, unlike the body which is made of parts, unlike everything else which can be decomposed into more and more elementary units. Conscious states cannot be reduced to constituent parts.

How did consciousness come to exist in the first place? Did it evolve from non-conscious properties? In that case, why? What purpose does it serve?

Could I be conscious of things that I am not conscious of? Am I in control of my consciousness? Is this conscious thought of mine only one of the many possible conscious thoughts that I could have now, or is it the only conscious thought that I could possibly have now? Is consciousness in control of me? This question is crucial to understanding whether there is a locus of consciousness in the brain, or whether consciousness is simply a side-effect of processes that occur in the brain.

The most frustrating property of consciousness is probably its opacity: we cannot know who and what is conscious. How widespread is consciousness? Who else is conscious besides me? Are other people conscious the same way i am? Are some people more conscious and others less conscious? Are some animals also conscious? Are all animals conscious? Are plants conscious? Can non-living matter also be conscious? Is everything conscious?

Can things inside conscious things be conscious? Are planets and galaxies conscious? Are arms and legs conscious?

What is the self? The self seems to represent a sense of unity, of spatial and temporal unity: "my" self groups all the feelings related to my body, and it also groups all those feelings that occurred in the past. My body changed over the years, and my brain too. All the cells of the body change within seven years. Therefore my "mind" must have changed too. But the self somehow bestows unity on that continuously changing entity. If we consider that our bodies are ultimately made of elementary particles, and that the average lifetime of most elementary particles is a fraction of a second, we can say that our bodies are continuously rebuilt every second. The matter of our bodies changes all the time. The only thing that is preserved is the pattern of matter. And even that pattern changes slowly as we age. Not even the pattern is preserved accurately. What makes us think that we are still the same person? How can I still be myself?

Laws that protect animals are not clear about "what" makes an animal worthy of protecting: killing a neighborhood cat because I don't like it is generally considered offensive, but killing a spider because I don't like it is absolutely normal. One can own a dog and file a suit against somebody who killed it, but one cannot own an ant and file a suit against somebody who stepped over it. Why slaughtering cows by the millions is a lawful practice and killing a pigeon in a square is a crime? We grant more rights to dog breeds (that are the result of genetic experiments performed mainly in Britain over the last two centuries) than to rats (whose brain we routinely use to understand our own brain because they are so similar). In fact, we are more likely to be opposed to killing a butterfly than a rat, simply because butterflies look beautiful, even if butterflies have the most primitive of brains.

The US physicist Erich Harth focused on the following properties of consciousness: "selectivity" (only a few neural processes are conscious); "exclusivity" (only one perception at the time can be conscious); "chaining" (one conscious thought leads to another one"); "unitarity" (the sense of self).

These properties of consciousness (partiality, sequentiality, irreducibility, unity, opacity, etc) set consciousness apart from any other natural phenomenon. And make it difficult, if not impossible, to study it with the traditional tools of the physical sciences.

Two Levels of Consciousness

We can take consciousness as a primitive concept (just like "time", "space" and "matter"), that we all "know" even though we cannot define it. We can define what the brain (or at least the neural system) is and what brain processes are. We can define cognition, as the set of cognitive faculties (learning, memory, language, etc), each of way is relatively easy to define.

When we refer to cognition, we are often interested in more than just the neural process underlying a cognitive faculty. We are interested in general questions such as "how can a living thing remember something" and "how can a living thing learn something". Such questions have two parts. The first part is about the mechanism that allows a piece of living matter to remember or learn something in the sense of being able to perform future actions based on it. The second part is about the awareness of remembering or learning something. The first part doesn't really require consciousness, and it may well be explained on a purely material basis. Even non-conscious things (non-living matter) may be able to remember and learn. Ultimately, the first part of the cognitive process can be summarized as: "Matter modifies itself based on occurrences in the environment so that its future behavior will be different". Fascinating and intriguing, but far less mysterious than the other half of the phenomenon: "... and, in the process, it is also aware of all of this".

The mechanisms that preside over memory, learning, language and reasoning can be described in material terms. And machines have been built that mimic those processes. The other half of the problem is still as mysterious as it was centuries ago. How does a brain process give rise to the awareness that the process is going on?

It looks like by "mind" we always meant something physical, material, reducible to physical processes inside the brain, which could be reproduced in a laboratory, and possibly on beings made of a different substance. But at the same time we also meant something that today's sciences cannot replicate in a laboratory: the awareness of that physical process going on inside us.

"Mind" encompasses both the cognitive processes (of memory and learning, language and reasoning) and the "feeling" associated with those processes: consciousness.

At closer inspection, "consciousness" is a term that encompasses a number of phenomena: thought, the self (the sense of the "i", the awareness of being), bodily sensations (such as pain and the color red), emotions (anger, happiness, fear, love). But not necessarily cognition (reasoning, memory, learning, etc).

There is a "narrative", "cognitive", "higher-level" consciousness, which is relatively detached from our bodily experience and which seems to rely on language, and there is an "experiential", "sensorial" consciousness, which has to do with sensations received from the senses, i.e. with our immediate bodily experience. The latter may be common to many species, while the former might be an exclusive of humans because it may require some additional level of circuitry in the brain than basic sensations or emotions.

The former is what we call "thought", including the self. The latter consist of "sensations" and "emotions".

Consciousness is the awareness of existing. Self is the awareness of lasting in space and time (of being an "i"). Sensations are bodily feelings such as pain, the color red, warmth. Emotions are non-bodily feelings such as anger, happiness, fear. Cognition encompasses the processes of reasoning, memory, learning, speaking, etc. Perception is the physical process of perceiving the world. Thought is the act of being conscious over an extended period of time.

Qualia

The problem of phenomenal qualities has puzzled philosophers for centuries. There is no "red" around me, just particles. Where does the "red" that I see come from? That red exists in my mind, but it does not exist outside my mind. Red is in me, not in the world.

How can a reality made of atoms of finite size be generating my feeling of something as uniform and continuous as the color red?

In 1929 the US philosopher Clarence Lewis called them "qualia" (from the Latin "quale", which is the dual of "quantum" and refers to the subjective aspect of a thing). Qualia are qualities that are subjective, directly perceived and known in an absolute way. The taste of something, the color of something, a pain or a desire are associated to qualia, to "feelings" of those things.

Qualia are subjective: I cannot be sure that another person's "red" is identical to my red.

Qualia are known in an absolute way: in another world red could correspond to a different frequency of light, and we would have to change the branch of Physics that deals with colors, but what I see as "red" I would still see as red.

Why does Nature present itself to my senses in two contradictory ways? If I believe my immediate perceptions, there is red. If I try to make sense of my perceptions, I work out a theory of Nature according to which there is no red, but only a vast mass of floating particles.

As a matter of fact, matter is "inscrutable" to our consciousness. We would like to think that, if nothing else, we know what the world is. We may be puzzled by the nature of mind, but we do know what matter is. At closer inspection, even matter turns out to be a bit of a mystery. We cannot perceive, and therefore conceive, what matter ultimately is. Our mind presents us with a game of illusions, whereby the world is populated with objects, and objects have shapes and colors. Science, on the other hand, tells us that there are only particles and waves. We cannot perceive that ultimate reality of matter. Matter is inscrutable to us.

We know what consciousness is because we feel it. We know what matter is because we sense it. Because we can sense it, we can build scientific theories on the nature of matter. But we cannot feel it, we cannot feel what matter ultimately is. Because we can only feel it, we cannot build scientific theories on the nature of consciousness. Although we can feel it.

Feelings Are Not in the Head

The US philosopher Michael Tye believes that our feelings are not in the head at all. Neurologists can never explain what it is like to smell or taste.

The starting point is a thought experiment by the Australian philosopher Frank Jackson ("Epiphenomenal Qualia", 1982). Imagine a scientist who knows everything about a subject, but has not experienced that subject. She has lived her entire life in a black and white environment but studied all there is to know about colors. She has seen colored objects only on a black and white television set. She just has not seen them in color. But she knows what color is and what properties it obeys and so forth. Then one day she steps outside her black and white environment and experiences the colors of those objects. No matter how much she knew about colors, when she actually sees a red object, she will experience something that she had not experienced before, she will "learn" something that she did not know: the "what it is like" of seeing a color (what Tye calls the "phenomenal character" of seeing a color). There is a difference between objective knowledge of something and subjective experience of something. The latter constitutes the "phenomenal consciousness" of something.

The British philosopher Bertrand Russell had already argued that light is precisely what a blind man cannot see. We can explain the theory of electromagnetic waves to a blind man, but light is precisely the thing that a blind man can never understand.

Tye believes that phenomenal states cannot be possibly realized only by neural states (as opposed to what physicalism claims). Tye believes that mental states are symbolic representations, but he differs from Fodor in that he does not believe that the representation for a sensation involves a sentence in the language of thought. The belief of something is represented by a symbolic structure which is a sentence. The sensation of something, instead, is represented by a symbolic structure which is not a sentence. The format (the symbolic structure) of a sensory representation is instead map-like: a pattern of activation occurring in a three-dimensional array of cells each containing a symbol and to which descriptive labels are attached. The patterns are analyzed by computational routines that are capable of extracting information and then attaching the appropriate descriptive labels.

A sentence would not be enough to represent a sensation, as a sensation includes some kind of "mapping" of the domain it refers to. For example, pain is about the body, and needs a way to represent the body parts that are affected by pain. Sentences lack this map-like representational power. Tye's patterns of activation in those map-like structures are therefore representations of bodily changes that trigger some computational processing. And this is what an emotion is, according to Tye.

Tye's hypothesis is that phenomenal consciousness is not in the neurons: phenomenal consciousness is in the "representations".

Tye believes that the body is equipped (as a product of evolution) with a set of specialized sensory modules for bodily sensations (for pain, hunger, and so forth) just like the specialized sensory modules for the five senses (physically different neural regions). Each module is capable of some computation on some symbolic structure.

Additionally, Tye notes that the object of a feeling is non-conceptual. We have different feelings for different shades of red even if we don't have different concepts for those shades of red. Thus we are capable of many more feelings than concepts.

Tye concludes that "phenomenal states lie at the interface of the non-conceptual and conceptual domains", at the border between the sensory modules and the cognitive system.

Tye analyzes the "phenomenal character" ("what it is like") of an experience and its "phenomenal content" ("what is being experienced"). Tye shows that the phenomenal character of an experience is identical to its phenomenal content: the feeling of pain in a foot cannot be abstracted and remains the fact that it is pain in that foot. Tye, therefore, concludes that phenomenal aspects are a subset of the representational aspects, and not distinct from them.

Because phenomenal character (the "what it is like" feeling) is phenomenal content, experiencing "what it is like" depends on having the appropriate system of concepts: one must have the appropriate system of concepts in order to understand what it is like to experience something. I cannot know what it feels like to be a bat because I don't have the appropriate concepts to feel what a bat feels.

Appropriate concepts are "predicative" and "indexical", which can be acquired only from direct experience (past or present, respectively).

Tye does not truly solve the "explanatory gap" between phenomenal states and physical states (how subjective feelings arises from neural states that are not subjective). His theory offers an explanation for why we cannot know "what it feels like" to be a bat, but does not explain why the bat feels whatever it feels, i.e. how feelings are created from brain states.

An Impossible Science?

It was the US psychologist Karl Lashley who first warned that... the mind is never conscious. The mind can never perceive the processing that goes on in the brain when the mind is thinking something. When I think about myself, I am not conscious of what my brain is doing. Whatever it is that I am feeling, it is not what the brain is doing. I am not aware of the billions of electrochemical processes switching neurons on and off.

One can even suspect that it is simply impossible for a conscious being to understand what consciousness is. The US philosopher Thomas Nagel pointed out that one is only capable of conceiving things as they appear to her, but never as they are in themselves. We can only experience how it feels to be ourselves. We can never experience how it feels to be something else, for the simple reason that we are not something else. As Nagel wrote ("What is it like to be a bat", 1974), we can learn all about the brain mechanisms of a bat's sonar system but we will never have the slightest idea of what it is like to have the sonar experiences of a bat. Likewise, understanding how the human brain works may not be enough to understand human consciousness.

Cognitive Closure

The British philosopher Colin McGinn argued that consciousness cannot be understood by beings with minds like ours. McGinn believes that there is nothing "magic" about consciousness: consciousness is a natural phenomenon just like many others (lightning or hurricanes or comets) and, as such, it is a consequence of the way matter is structured and functions (specifically, how the brain works).

We are capable of understanding natural phenomena like lightning and hurricanes, but McGinn suspects that we are not capable of understanding "all" natural phenomena. There are natural phenomena that our mind cannot comprehend, just because our mind is not an infinitely powerful computer.

In a sense, McGinn's central thesis is that our mind has limitations. Consciousness itself might be one of the phenomena that fall within the mind's limitations, i.e. fall outside the "cognitive closure" of the human mind. That does not mean that nobody can ever explain consciousness: a being equipped with a "better" mind could understand how consciousness works, where it comes from and what it is. But not our mind.

We can understand how the brain works. The brain is a natural phenomenon that we can easily investigate with our science. We will learn more and more about the brain. We will eventually work out a very detailed model of the brain. We may even be able to reproduce the brain molecule by molecule. But we will never be able to figure out how emotions arise from the unconscious matter that makes up the brain.

McGinn's fundamental assumption is that the human mind is biased in its cognitive skills. This follows from a Darwinian view of life: all of our organs are biased, one way or another, towards coping with the environment. McGinn simply extends this principle to the mind. Our mind is very skilled at understanding spatial and temporal relationships, and at doing what we call Science. Our mind is probably not very skilled at doing things that would be useful on Mars but that do not exist on Earth. Thus it is reasonable to assume that the human mind has been designed by evolution to solve some problems better than others, and not to solve many other problems at all.

In particular, McGinn thinks that our intelligence is not designed to understand consciousness. Science is the systematic understanding of nature by the human mind, but it is limited to what the human mind can understand. There might be many things in nature that the human mind will never understand, and maybe not even perceive. Consciousness is one of them. Our brains were not biologically designed to understand consciousness. McGinn even speculates that knowledge of ourselves is useful to a limit: maybe if we could fully understand ourselves, we would get very depressed and not willing to survive anymore. Thus natural selection may have pruned away the ones who did understand consciousness, and left only the ones who could not understand it, of whom we are the descendants.

Inspired partly by the British philosopher Bertrand Russell and partly by the German philosopher Immanuel Kant, McGinn argues that consciousness is known by the faculty of introspection, as opposed to the physical world, which is known by the faculty of perception. The relationship between one and the other, which is the relationship between consciousness and brain, is "noumenal", or impossible to understand: it is provided by a lower level of consciousness that is not accessible to introspection. That is why consciousness does not belong to the "cognitive closure" of the human organism.

Understanding our consciousness is beyond our cognitive capacities, just like a child cannot grasp social concepts or i cannot relate to a farmer's fear of tornadoes. McGinn points out that other creatures in nature lack the capacity to understand things that we understand (for example, the general theory of relativity). There are parts of nature that they cannot understand. We are also creatures of nature, and there is no reason to exclude that we also lack the capacity to understand something of nature. We may not have the power to understand everything, unlike what we often assume. Some explanations (such as where the universe comes from, and what will happen afterwards, and what is time and so forth) may just be beyond our mind's capacity. Explanations for these phenomena may just be "cognitively closed" to us. Phenomenal consciousness may be one such phenomenon.

"Mind may just not be big enough to understand mind".

McGinn speculates that consciousness might be a very ancient invention. Indeed, the fact that consciousness has no spatial dimensions leads him to speculate that consciousness may have tapped into a non-spatial property that disappeared with the Big Bang (the cosmic event that created the spatial universe we live in). Our minds may be remnants of a dimension that does not exist anymore but that was pervasive in the pre-Big Bang universe.

He cunningly refutes the idea that computers can be conscious in virtue of being computers. McGinn explains that this idea is based on a bad theory of consciousness. We have no evidence that the property of running a program is the property that yields consciousness. This does not mean that conscious machines are impossible: the key is in finding out what is the property that yields consciousness, and then implementing that feature in a machine. McGinn points out that one such machine already exists: me. Thus it is feasible. Cogito, ergo I am feasible. McGinn changes the question to make it more interesting: can a machine made of inorganic material be conscious?

Implicitly, McGinn assumes that our "cognitive closure" does not change with time, that it is a constant of the human condition. On the other hand, it is obvious that it changes during the course of a lifetime: children cannot grasp concepts that adults can. One can also argue that our "cognitive closure" has evolved over the centuries, that we are more "conscious" today than we were thousands of years ago. Certainly, concepts such as democracy and women's rights are more obvious to today's humans than they were to even the most enlightened of the ancient Greek philosophers. Studies on ancient texts point to a reliance on gods that today has been replaced by a reliance on our own opinions. One can argue that today we "think" differently. One can argue that each generation uses knowledge from the previous generation to expand that "cognitive closure".

The question then is whether the "cognitive closure" that McGinn talks about is a temporary limitation, a "stage" in the evolution of manking, or a permanent deficiency of our mind (due, say, to the structure of the brain, or to some impossible neural connections, or to the limited capacity of our memory). Since McGinn does not "quantify" where the cognitive closure comes from, i.e. what it is physically, one cannot decide whether it can be overcome or not by future generations. If we do not know what causes that cognitive closure, we cannot know whether it can be overcome.

The Kenyan-born biologist Richard Dawkins, following the British biologist John Haldane, believes that, just like there are limits to what human eyes can see because they only see what made sense to evolve over millions of years, there is no reason to believe that evolution shaped the human brain to perform any more function than what is needed to survive in its environment. For example, we can only see in black and white mode at night, and we don't see (nor hear) the extreme parts of the spectrum of frequencies. If that is true for the eyes, it is likely to be true also for the brain: the human brain evolved the ability to "think" what matters for the survival of humans, but not the ability to "think" other thoughts, which are therefore beyond the ability of the human brain. Just like we cannot run as fast as a cheetah, we cannot think as profoundly as some other being that (who?) has evolved or will evolve a better kind of brain.

Nonetheless, one can object: if our brains were not designed to understand consciousness, why are we wondering about it? Were our brains designed to wonder about consciousness? Were our brains designed to wonder about something that we cannot possibly understand? Why?

Infinite Knowledge

And so the Israeli physicist David Deutsch counters that we used to live in a world of supernatural entities. Then we discovered that we are capable of explaining how the world works because the world obeys mathematical laws that our brain can master. Therefore we became universal beings, capable of understanding anything that happens in the universe ruled by those laws of nature. A higher intelligence (incomprehensible to us) can only exist if the universe is not explicable, i.e. if there exist supernatural beings. In other words, the cognitive closure of human brains would imply that the ancient superstitions were correct.

It is the scientific "explanation" that allows us to create a long chain of interpretation that allows us to understand phenomena that are very far from our everyday experience. For example, astronomical observation consists of, in practice, just looking into devices built out of humble materials found here on Earth but, ultimately, allows the mind to visit distant worlds. The mind can "leave" the body for this spectacular voyage because of the "explanations" that guarantee it is not just being tricked by drugs. There is something special about this process that makes humans truly unique.

Deutsch disagrees with those who think that there is nothing special with this typical planet of a typical star of a typical galaxy (the principle of mediocrity): humans have created conditions (for example, low-temperature refrigerators) that are very rare in the universe. Deutsch emphasizes that most of the universe is very different from the environment that humans created in their cities, homes and especially laboratories. What made humans unique is that they can use technology to expand the capabilities of their brains. Whatever caused their brains to evolve the way they did, those brains are now capable of going beyond their original survival functions. And the reason that technology does that is the explanatory knowledge that only humans have learned to master. What is unique about explanatory knowledge is obvious if one considers that humans managed to survive in an extremely hostile biosphere and even moved to environments that are not the ones in which they were shaped by natural selection. Other species have knowledge too, but the kind of straightforward knowledge that limits their reach to their habitat. The explanatory knowledge of humans allows them to do things that their brains were not programmed for and their bodies could not do without help from technology.

Thus Deutsch calls humans "universal constructors": humans can transform anything into anything as long as they understand the natural laws that govern the universe and then build technology according to those laws. If the universe is governed by deterministic rules, then the human brain can eventually (explanation after explanation) get to understand everything. We are capable of everything, and in particular of understanding everything. There is nothing that obeys the laws of nature that we cannot understand. If such a thing existed, it would have to be "supernatural" by definition: not obeying the laws of nature. In a universe governed by mathematical laws, the reach of explanatory knowledge is infinite, and therefore so is the reach of a brain that is capable of acquiring explanatory knowledge recursively (one explanation leading to another one). There is no limit to human knowledge. There is no limit to human creation.

Historically, it wasn't always this way. At some point humans became what their brain made possible: universal constructors. He links this conceptual revolution with the Enlightenment, that spawned rebellion against authority in many fields, including science: those rebels looked for explanations, not just dogmas. Furthermore, the Enlightenment introduced the notion that progress is good and should be a universal ideal to achieve by society. So the chain of explanations that leads to more and more knowledge became not only possible but an explicit goal of humankind.

The effect of brains like ours (brains that can deal with explanatory knowledge and therefore become universal constructors who can alter their environment at will) is significant. Deutsch points out that astrophysics is incomplete without a theory of people because people can alter the course of events that the laws of nature alone would cause. If we don't consider the actions of people, we can't understand why the Earth is the way it is.

Furthermore, the effect of a knowledge-processing brain is profound also because an intelligent deliberate transformation is actually more likely to happen than a spontaneous one: the chances that an object is changed by a spontaneous transformation (one caused only by the laws of nature) are relatively low in the grand scheme of things, whereas the chances that an object is changed by an intelligent being for whatever purpose are very high.

In other words: the universality of the laws of nature coupled with a universal constructor like the human mind yields unlimited knowledge growth.

Deutsch argues that the well-known limitations on mathematics and computations (Godel's Theorem and the likes) do not affect his claims that humans can extend knowledge indefinitely: one can understand a mathematical statement without proving it. Basically, the proof is a technicality. If you can't prove it, it doesn't mean you don't "know" something. The only limit to human knowledge that Deutsch sees is the impossibility to predict future knowledge: the fact that we can find solutions to any problem does not mean that we already know those solutions.

The physics of knowledge representation is that the physical system of the brain somehow contains a model of the physical system of the entire universe. As more and more knowledge is acquired that model becomes a better and better approximation of the universe (and viceversa). The world is knowledge-friendly because it obeys the laws of nature. The human brain is knowledge-capable. Starting with the Enlightenment, the human brain decided to exploit this capability. The spirit of the Enlightenment provided the motivation that was missing. Knowledge allows the brain to build technology to help acquire more knowledge even about facts that the brain cannot experience directly. This chain reaction leads to an endless creation of knowledge.

Deutsch's argument that, assuming the universe is run by deterministic regular "laws of nature", a mind that can understand those laws of nature can in principle understand everything, is persuasive. However, he doesn't talk about the one feature of this universe that no science has been able to express, describe and predict with mathematical laws: human consciousness; which also happens to be the one that i'm sure about, and the one i care most about. Colin McGinn has argued convincingly that the human mind just might not be capable of understanding it, no matter what. Deutsch provides a general framework in which the human mind can understand everything that can be expressed in mathematical laws, but the doubt remains that some things will never be expressed in a formal, mathematical form.

Non-Event Causation

The US philosopher John Searle makes a two-fold claim: consciousness cannot be reduced to the neurological processes that cause it, but is indeed a biological feature of the brain.

Brains cause minds, in his opinion, although we will not find feelings and emotions in the material processes of the brain, because feelings and emotions are higher-order features of the brain. Searle attacks the Cartesian tradition from the foundations: both dualism and materialism make no sense. The division of the world into matter and mind is arbitrary and counterproductive. In his view, we simply have to face the facts: consciousness is caused by brain processes, but consciousness cannot be reduced to those brain processes because it is a "first person" phenomenon and the brain processes are "third person" phenomena. To Searle, the mind-body problem has never existed: René Descartes invented a vocabulary, a terminology, not a real problem.

In a similar vein to the school of "supervenience", John Searle compares the mind-body problem to explaining how electricity arises from electrons or liquidity from molecules.

Searle is content with stating that consciousness is a (causally) "emergent" property of systems, just like electricity and liquidity. Searle realizes that liquidity can be predicted from the properties of elementary particles, whereas consciousness cannot be predicted from the properties of neurons. Searle realizes that Physics can explain how the features of electricity correspond to the features of electrons, whereas we can't explain (yet) how the features of consciousness arise from the features of neurons. Physicists can explain why (and exactly under which conditions) a set of molecules can achieve the phase transition to liquidity, whereas neurologists can't explain when exactly non-conscious matter becomes conscious. Searle thinks that it is not just a limit of today's neurophysiology (likely to change with time), but that this will always be the case, that it is impossible to provide a material explanation of the features of consciousness. Searle thus admits a crucial difference between consciousness and electricity or liquidity or digestion: consciousness is special in that it cannot be explained.

Searle has nothing to offer other than declare that mental life exists and that it emerges from neurons, which is equivalent to saying that liquidity exists and that it emerges from liquids.

Searle thinks that computers have no minds because they are not brains, but he never proves the underlying assumption: that they are not brains. In Searle's jargon, "brain" is simply the "thing" that enables the mind. His entire theory can therefore be viewed as a mere tautology: the mind is due to the thing that causes it. What that thing is remains a mystery. Ultimately, Searle merely states that the mind exists. If one has not defined what a brain is, it is hard to claim that something is not a brain.

Searle basically resurrects Thomas Nagel's argument that consciousness cannot be explained. Thus, consciousness is not an emergent process like liquidity, because liquidity, like all emergent properties, is reducible to the physical process that creates it. Emergent properties are normally predictable by science: we know when (and why and how) a substance is a liquid and not a solid or a gas. If consciousness is indeed an emergent property, why should it be the only one that we cannot predict and explain?

Searle is not baffled by the emergence of conscious feelings from unconscious neurological processes of the brain. He finds it perfectly understandable. And therefore he downplays and ridicules all theories that tried to solve this paradox. But it is like somebody not being puzzled by the fact that the sun rises and sets every day, and contenting himself with the idea that it must be a feature of the Earth.

The Classical Theory of Consciousness

William James (at the end of the 19th century) was responsible for articulating the "classical" theory of consciousness, the equivalent of Isaac Newton's classical Physics. To James, consciousness is a sequence of conscious mental states, each state being the experience of some content. Just like Newton saw a unitary and continuous space, James saw a unitary and continuous consciousness.

James thought that consciousness must have an evolutionary purpose, just like Darwin thought that all features of the body must have an evolutionary purpose. Thinking is useful for our survival, just like eating and mating. James treated consciousness like a function, not an entity.

James was, in part, reacting to the theory of perception that dated from the 19th-century German physicist Hermann von Helmholtz, that sense data from the senses are turned by the "mind" into percepts which are conscious experiences of the environment. James thought, instead, that the output of brain processes is guidance of action in the environment, not a conscious experience of the environment.

Furthermore, James realized that every act of perception specifies both a perceiving self and a perceived object. Seeing something is not only seeing that object: it is also seeing it from a certain perspective. The perspective implies a "seer". A sensory act specifies not only the environment but also the self. Self (the "subjective") and the environment (the "objective") are two poles of attention. Each act of perception specifies both the self and the environment.

The self, in turn, is one but is also divided. The self is partly object and partly subject: there is a self who is the knower (the "i") and a self who is the known (the "me").

Consciousness in the Environment

In the 1930s the US biologist George Herbert Mead put forward a theory that, instead, located consciousness in the world, outside the organism.

For Mead, consciousness is not a separate substance, but the world in its relationship with the organism. Objects of the environment are colored, beautiful, etc: that "is" consciousness.

Objects do not exist per se: they are just the way an organism perceives the environment. And, presumably, different organisms may perceive different objects. Each organism perceives a different environment. It is our acting in the environment that determines how we perceive the environment. We are actors as well as observers. The response of the environment is what we perceive as objects.

In other words: the nature of the environment lies in its relationship with the organism; the environment results from the actions of the organism, in response to the stimulation of its sense organs, i.e. the organism determines the environment; and this results in the appearance of objects.

We are programmed to pick up organized information about the environment in the form of objects. The environment we perceive is merely a perspective of the real environment (that Mead calls "habitat"), one of the many that are possible. Each organism gets a different environment from the same habitat. The one world that we living beings inhabit is perceived in different ways (as different environments) by different organisms. This perspective is determined by the actions that the organism is capable of, by the set of its "organized responses". In particular, any change of the organism results in a change of the environment.

The organized objects of the environment actually represent our organized responses. They are what exists for us (for the kind of responses we are capable of). There is a direct relationship between our repertory of actions and our view of our habitat. Ultimately, the environment is a property of the organism as well as of the habitat. Those objects have qualities (such as colors) and values (such as beauty) that constitute what we call "consciousness". But, again, those objects and therefore their qualities depend on our repertory of actions.

Consciousness is a function, not a substance, and it refers to both the organism and the environment. It is located in the organism's environment, not in the organism's brain.

Consciousness is not a brain process: the switch that turns consciousness on or off is a brain process. Pulling up the blinds of a window does not create the street, it merely reveals it. By the same token, the brain can "pull up the blinds" and reveal consciousness, i.e. the set of objects and their qualities. Or it can pull them down, and consciousness disappears. The brain only has control over this switch.

Everybody has this kind of consciousness, but some species (and the children of our own species) cannot report on their experiences. It is social experience that makes awareness possible. It is not consciousness that enables socialization: it is socialization that enables consciousness (as awareness of one's experience).

Mead speculated that the individual views society as the "generalized other". The self arises from belonging to a group. The individual plays a role within the group. In fact, usually the individual belongs to many groups, and therefore plays many roles. Each role contributes to shape her or his self. The individual gradually integrates all these roles in one comprehensive view of her or his self, which is equivalent to saying that the individual takes the viewpoint of the "generalized other". More accurately, this constitutes the "me". Mead believed that we also have an "i", that is subjective and not socially constructed.

The self is created through socializing. A self is a contributor to awareness, and is aware of all contributions. A self always belongs to a society of selves. And a self is what she is as a member of that society of selves.

Likewise, the US biologist James Jerome Gibson used vision to explain what awareness is and what it is not. Body and mind constitute a false dichotomy. Awareness is both physical and mental. Awareness is a function performed by a living observer, the whole living being, not just its mind or its body. Awareness is a biological phenomenon. Perceiving is keeping in touch with the world. The observer is not external to the world, and therefore her/his awareness cannot be a state outside the world (i.e., in a different substance called "mind"). Cognition is a biological phenomenon, and it is both mind and body. Awareness is both mind and body.

Gregory Bateson believed that minds extended beyond bodies, that consciousness was not only in the brain but also in the surrounding environment.

The Czech psychologist Staninslav Grof thinks that we are not the makers of consciousness but merely the transmitters of it.

Sensing the Brain

Several models of consciousness focus on its behavior as a "sense" capable of perceiving the processing of the brain.

The British philosopher Nicholas Humphrey argued that to be conscious is to feel sensations, as opposed to perceptions. Animals have developed two ways of representing the interaction between the body and the world: "affect-laden" sensations and "affect-neutral" perceptions. Sensation and perception are separate and parallel forms of representation. Sensations are to be found at the boundary between the organism and the world and at the boundary of past and future. One "senses" a circle of light hitting the retina; one "perceives" the sun in the sky. One can have sensations about perceptions and perceptions about sensations.

Humphrey ("The Use of Consciousness", 1967) speculated that we possess an "inner eye" that behaves like any other sense, except its object is the brain itself. This inner eye evolved because it was useful, and it evolved in such a way as to be useful. Therefore it does not deliver a one-to-one picture of the brain's activity, but a selected, abridged and biased one, as useful to survival as possible. Consciousness allows me to perceive the state of my brain as conscious states.

The US philosopher William Lycan is a proponent of consciousness as "internal monitoring", an idea that reaches back to the British philosopher John Locke and the German philosopher Immanuel Kant. Lycan emphasizes that most of our mental life is unconscious, that we are conscious of only part of it. A possible explanation is that consciousness is a perception of our own psychological state, of what is going on in the mind (Kant's "inner sense").

The British neurophysiologist John Eccles believes that consciousness resides in a psychological world that transcends the physical. Eccles believes that the mind is an independent entity that exercises a controlling role upon the neural events of the brain by virtue of its interaction across the interface between World 1 (the physical world) and World 2 (the mental world). The mind is continuously searching for brain events that are interesting for its goals.

The US linguist Ray Jackendoff believes that there are three main entities that account for our mental life: the physical brain, the computational mind (cognition) and the phenomenological mind (consciousness). The computational mind is the one that really "thinks", whereas the phenomenological mind only "feels" superficially a subset of the "thoughts". Most of "thinking" is actually unconscious. We are never conscious of the outer world, but only of the shadows of some of the processing that the computational mind does on the outer world. Consciousness arises from a level of representation that is intermediate between the sense-data and the form of thought, at the border between the representations of the inner and of the outer worlds.

The US psychologist Owen Flanagan does not believe in "one" consciousness, but in a group of "conscious" phenomena. Some of the processes of our body are unconscious and non-perceived (e.g., the heartbeat), while some are unconscious but perceived by other processes (sensors), and some are conscious, perceived by themselves. Consciousness is a heterogeneous set of processes, not a substance or an object. Flanagan's theory is, de facto, a variation on William James' stream of consciousness: there is no "Mind's i" that thinks and is conscious, there are just thoughts that flow. Consciousness is not: consciousness flows. "The thoughts themselves are the thinkers". I do not think thoughts, thoughts think me. There is no "i": there is conscious activity. This conscious activity differs from the neural activity only in kind, but the mind "is" the brain.

Binding Brain and Consciousness

Knowledge about the world is distributed around the brain. How it is then integrated into one unitary perception is the "binding" problem.

This problem occurs at several levels. A sensory input is channeled through several different areas of the brain, each brain region focusing on one aspect of the input, but then somehow the mind perceives the whole input as something that happens at the same time in the same place, and it is a whole. The "binding" problem refers to how the brain creates the whole perception out of a sensory input that has been fragmented around the brain.

For example, a visual input is "split" so that one brain region analyzes the shape and one brain region analyzes the color. But somehow these separate pieces of information are joined again to produce the overall sensation of the image. At a higher level, different sensory inputs come together: the sound of an event is merged with the image of the event, or the smell of the event, or the touch of the event. The result is the overall feeling of the situation.

At an even higher level, the situation is merged with pre-existing memories and concepts. We don't only see a human being moving and speaking around us: we see our friend X talking to us. At the highest level, this entire complex system of feelings and knowledge "feels" unified in our consciousness. There is "one" feeling of "me" existing in a "world". Somehow all has been "bound" together into consciousness.

There are different theories about where and how and when this ultimate form of "binding" could occur.

"Space-based binding" is advocated by scientists who believe that there is a specific place in the brain where all information is integrated together. In the 1990s, a competing paradigm has emerged which is based on time instead of space, and is therefore referred to as "time-based binding": there is no particular place where the integration occurs, because integration occurs over the entire brain, and is regulated by some periodic process.

Space-based binding theories try to identify the "homunculus" in the brain that is responsible for running the integration process.

The working memory is a popular candidate for such a task, but no piece of the brain seems likely to show us the transformation of electrochemical processes into "feelings" (conscious processes).

According to the Portuguese neurobiologist Antonio Damasio, the story is more complex. There is not just one working memory: there is a whole system of "convergence zones". The brain has "convergence zones" and convergence zones are organized in a hierarchy: lower convergence zones pass information to higher convergence zones. Lower zones select relevant details from sensory information and send summaries to higher zones, which successively refine and integrate the information. In order to be conscious of something, a higher convergence zone must retrieve from the lower convergence zones all the sensory fragments that are related to that something. Consciousness of something occurs when the higher convergence zones fire signals back to lower convergence zones.

The Movie in The Mind

Damasio breaks the problem of consciousness into two parts: the "movie in the brain" kind of experience (how a number of sensory inputs are transformed into the continuous flow of sensations of the mind) and the self (how the sense of "owning" that movie comes to be).

The "core" consciousness of the "movie in the brain" is essentially unchanged throughout a lifetime, and humans share it with many other species.

On the other hand, the "extended" consciousness of the self is refined over a lifetime: an "owner" and "observer" of the movie is created within the core consciousness, in such a way that it seems to be located outside the brain, while it is part of the brain's neural processes and part of the movie itself which those neural processes generate. The more developed the sense of the self, the stronger the impression that the movie in the mind is "my" experience of the world.

Distinct parts of the brain work in concert to represent sensory input. Brain cells represent events occurring somewhere else in the body. Brain cells are "intentional" (the philosophical "intendo"): they represent something else in the body. They are not only "maps" of the body: besides the topography, they also represent what is taking place in that topography.

Indirectly, the brain also represents whatever the organism is interacting with, since that interaction is affecting one or more organs (e.g., retina, tips of the fingers, ears), whose events are represented in brain cells.

These two "orders" of representation are crucial for the rise of consciousness.

The "movie in the mind" is a purely non-verbal process: language is not a prerequisite for this first level of consciousness. The "i" is a verbal process that arises from a second-order narrative capacity.

The brain stem and hypothalamus are the organs that regulate "life", that control the balance of chemical activity required for living, i.e. the body's homeostasis. Consequently, they also represent the continuity of the same organism.

Damasio believes that the self originates from those biological processes: the brain is equipped with both a representation of the body, and a representation of the objects the body is interacting with. Thus it can discriminate self and non-self, and generate a "second order narrative" in which the self is interacting with the non-self (the external world). This second-order representation occurs mainly in the thalamus.

More precisely, the neural basis for the self resides in the continuous reactivation of 1. An individual's past experience (which provides the individual's sense of identity) and 2. A representation of the individual's body (which provides the individual's sense of a whole). An important corollary is that the self is continuously reconstructed.

From an evolutionary perspective, we can presume that the sense of the self is useful to induce purposeful action based on the "movie in the mind". The self provides a survival advantage because the "movie in the mind" acquires a first-person character, i.e. it acquires a meaning for that first person, i.e. it highlights what is good and bad for that first person, a first person which happens to be the body of the organism, disguised as a self.

This second-order narrative derives from the first-order narrative constructed from the sensory mappings. In other words, all of this is happening while the "movie" is playing. The sense of the self is created, while the movie is playing, by the movie itself. The thinker is created by thought. The spectator of the movie is part of the movie.

Consciousness is an internal narrative, due to those mappings. The "i" is not telling the story: the "i" is created by stories being told in the brain ("You are the music while the music lasts").

Consciousness As Self-Reference

The idea of some form of "self-referential feedback" (of some kind of loop inside the brain) is firmly rooted in modern space-based binding theories. Gerald Edelman's "reentrant maps" and Nicholas Humphrey's "sensory reverberating feedback loop" are variations on the same theme. The idea is that, somehow, the brain refers to itself, and this self-referentiality, somehow, unchains consciousness. Rather than "space-based", these theories tend to be "process-based", since they are not only looking for the place where the binding occurs but also for the way it occurs, and the process turns out to be much more important than the place.

According to Edelman, consciousness is a natural development of the ability to build perceptual categories (such as "blue", "tall", "bird", "tree", "book"), the process that we normally call generalization. The brain can do this because neurons get organized by experience in maps, each neural map dealing with a feature of perceptions (color, shape, etc.).

First of all, Edelman distinguishes between primary consciousness (imagery and sensations, basically being aware of things in the world) and higher-order consciousness (language and self-awareness).

For primary consciousness to appear a few requirements must be met. It takes a memory, and an active type of memory, that does not simply store new information but also continuously reorganizes (or "re-categorizes") old information. Then it takes the ability to learn, but learning is not only memorizing, it is also a way to rank stimuli, to assign "value" to stimuli, to value one experience over another. A new value will typically result in a new behavior, and that is what learning is about. Then it takes the ability to make the distinction between the self from the rest of the world, i.e. a way to represent what is part of the organism and what is not. Then it takes a way to represent chronology, to order events in time. Finally, it takes a maze of "global reentrant pathways" (i.e., forms of neural transmission that let signals travel simultaneously in both directions) connecting all these anatomical structures. Primary consciousness arises from "reentrant loops" that interconnect "perceptual categorization" and "value-laden" memory ("instincts"). In general, cognitive functions emerge from reentrant processes.

Consciousness therefore arises from the interaction of two parts of the neural system that differ radically in their anatomical structure, evolution, organization and function: the one responsible for categorizing (external stimuli) and the other responsible for "instinctive" behavior (i.e., homeostatic control of behavior). Consciousness emerges as the product of an ongoing categorical comparison of the workings of those two kinds of nervous system.

From an evolutionary point of view, the milestone moment was when a link emerged between category and value, between those two different areas of the brain. That is when the basis for consciousness was laid.

A higher-level consciousness (being aware of itself), probably unique to humans, is possible if the brain is also capable of abstracting the relationship between the self and the non-self, and this can only happen through social interaction, and this leads naturally to the development of linguistic faculties. Edelman identifies the regions that are assigned to define self within a species (the amygdala, the hippocampus, the hypothalamus) and those that operate to define the non-self (the cortex, the thalamus and the cerebellum).

Note that, according to Edelman, concept-formation preceded language. Language was enabled by anatomical changes. What changed with the advent of language is that concepts became independent of time, i.e. permanent. And semantics preceded syntax: acquiring phonological capacities provided the means for linking the preexisting conceptual operations with the emerging lexical operations.

In Edelman's picture, consciousness is liberation from the present. Animals tend to live in the present, simply reacting to stimuli. Only conscious animals can think about the past and about the future.

As for the "place" where consciousness happens (its neural correlate), Edelman noted that consciousness is unified and a "whole", while nothing in the brain seems to be unified and a "whole": in fact, the brain is made of a multitude of regions that exhibit independent personalities. Consciousness must therefore be due to a global process that encompasses more than one region. He believes that the thalamocortical system originates such an activity: a massive, coherent (synchronized) activity by all regions of the brain, that transcends the individual activity of each region. Basically, consciousness is a process that happens throughout the brain, not manufactured in a specific region.

The location of consciousness is changing all the time, as different groups of interacting regions form and dissolve. At any moment in time, the "dynamic core" of primary consciousness is located in the interaction between the thalamus and the cortex. He does envision one particular region as being the permanent site of consciousness. In a sense, consciousness is the "process" not the "place".

The Mundane Components of Consciousness

British neurologist Adam Zeman showed how mundane consciousness is: a little more or less of this or that chemical makes a big difference as to how you "feel". Consciousness depends, mostly, on events that take place within the brain. A lack of this or that chemical is enough to alter our personality. After all, consciousness is a product of neural activity, and neural activity is a material process that uses material elements, which, ultimately, are the (indirect) constituents of consciousness.

By the same token, he describes perception as the brain's reaction to being bombarded with energy picked up by the body's sensors. Without that external energy, there would be no visual or auditory or any other kind of processing. Whatever the brain does, it is initiated by an energy impulse coming from the outside. It is then processed according to the chemical structure of the brain, which, by definition, depends on the amount and kinds of chemicals in the brain.

Zeman is quite convinced that the thalamus is the central site of consciousness. During sleep, the thalamus interacts with the cortex in rhythmic bursts, while inhibiting all sensory inputs. When the body is awake, the thalamus works as an intermediary between the periphery and the cortex, shuttling back and forth sensory inputs and commands to move. The brainstem is the switch that turns the thalamus on and off.

Zeman's experiments raise the issue of whether consciousness is really a whole that cannot be reduced to components. Different neurotransmitters seem to contribute to different aspects of consciousness. If a neurotransmitter is inhibited, the person is affected, both physically and emotionally. If each neurotransmitter helps shape consciousness, can't we also claim that consciousness is not a whole but, trivially, a sum of its parts?

Time-Based Binding

Time-based binding does not look for the "place" but for the "time" at which the binding of our conscious experience occurs. The general concept, that "waves" of neural activity in the brain account for the binding, was already envisioned by the US neurobiologist Walter Freeman in the 1970s.

The German neurologists Wolf Singer ("Stimulus-specific neuronal oscillations in orientation columns of cat visual cortex", 1989) and Christof Koch ("Collective Oscillations in the Visual Cortex", 1989) pointed out that at, any given moment, a very large number of neurons oscillate in synchrony, reflecting something that was captured by the senses, but only one pattern is amplified into a dominant 40 Hz oscillation. That is the "thing" the individual is conscious of. Out of so many oscillations of synchronized cells in the brain, one is special and happens to have a frequency of 40 Hz.

The British biologist Francis Crick originated the view that synchronized firing in the range of 40 Hertz in the areas connecting the thalamus and the cortex might explain consciousness; that consciousness arises from the continuous dialogue between thalamus and cortex. Awareness of something requires "attention" (being aware of one object rather than another), and attention requires "binding" (banding together all neurons that represent aspects of the object). Crick believes that binding occurs through a synchronous (or "correlated") firing of different regions of the brain. During attention, all the neurons that represent features of an object fire together. It is not just the frequency of firing of the neurons that matters, but the moments when they fire. The main difference between Llinas and Crick is in their background. Crick studied visual awareness and so is interested in consciousness that arises from external stimuli. Llinas, on the contrary, is more interested in consciousness that does not arise from external stimuli (what we call "thought").

Koch and Crick focused on the "neuronal correlates of consciousness" (NCC): the "thing" in the brain that corresponds to states of awareness. One of the first clues (although apparently disconcerting) is that much of the neural activity of the brain is not conscious at all: we are not aware of most of what our brain does. Most of the time (e.g., habits, instinct, etc.) the brain makes key decisions without "us" being conscious of those decisions.

Because most neuronal activity does not yield a state of awareness, they were led to believe that multiple forms of neuronal activity exist (this is almost a tautology). They speculated that one could potentially be conscious of many competing views, but only one "wins" the competition and results in awareness of the corresponding view.

All body cells are, to some extent, influenced by what happens to the body, but only a minority of the body cells represent external stimuli in an "explicit" manner: Koch and Crick believe that one is only conscious of features that are encoded "explicitly" by some neuronal assembly. The other way to find them is to listen to the way they oscillate. The electric potential of the brain as a whole exhibits oscillatory behavior in different frequency bands: the dominant rhythm for resting individuals is in the "alpha" band (8-12 Hz); the rhythm for normal cognitive activity is in the "beta" band (15-25 Hz) or, for more complex operations, in the "gamma" band (30 Hz or higher and typically 40 Hz); sleep is in the "delta" band (1-4 Hz). Each of these "oscillations" is caused by some synchronous behavior ("firing") of many neurons.

The "binding" problem is the problem of how the various features are integrated in the brain into the perception of the object as a whole, especially when the same brain is integrating other features of other objects. The German physicist Christoph von der Malsburg ("How are Nervous Structures Organized?", 1983) was the first to propose that synchronization could be the solution to the "binding" problem: the neurons working on one object are synchronized, and they are not synchronized with other populations of neurons that are working on other objects. There is one oscillatory behavior by neurons that seems to be associated with awareness, and it is in the 30-70 Hz range, with a peak around 40 Hz. Koch and Crick ("Towards a neurobiological theory of consciousness", 1990) claimed that this oscillation accounts for consciousness (i.e., that the set of those synchronized neurons "is" the NCC for the current state of awareness).

Koch and Crick believe that several such "coalitions" of neurons exist at every point in time, and a sort of Darwinian selection determines which one (and only one) wins and results into awareness.

Another clue to finding the NCC is the cholinergic system: consciousness only occurs when there is an adequate supply of acetylcholine neurotransmitters, which are regulated by the brainstem (people whose brainstem is damaged lose consciousness).

The brain has a convoluted structure, and the way it represents an experience is even more convoluted, but we perceive an experience as a sequence of events. Koch thinks this has to do with the fact that, at every point in time, only one coalition is the winning one. It may change all the time, but we perceive an ordered sequence of events, because every other coalition that is active at the time is suppressed. We do not perceive the convoluted activity of the brain, which is analyzing an overwhelming amount of data, but only those events that correspond to the winning coalition.

Koch divides short and long-term memory based on the underlying mechanism: long-term memory is caused by a physical rewiring of the brain (strengthening of connections), whereas short-term memory is caused by a sustained firing pattern by an assembly of neurons. Koch proves that consciousness depends on the latter, not on the former. Short-term (or, better, working) memory could provide a sort of "Turing test for consciousness": any being that displays a working memory is likely to be conscious.

Koch also speculated on why we are conscious at all. After all, we do not need consciousness: the brain makes most of the key decisions in an unconscious way. The autonomic system directs the organs to do their job, and "instinct" helps the body survive. Qualia (the qualitative aspects of things) are "symbols" that help the brain synthesize, summarize, huge amounts of data. Seeing red or feeling pain are shortcuts to handling huge amounts of sensory data. Qualia are symbols that summarize the state of the world (including the body itself). This is the "executive summary hypothesis".

Koch believes in a non-conscious homunculus, residing in the front of the forebrain, that handles the information stored in the back of the cortex (the sensory regions) and that does all of the "thinking". The front of the cortex is looking at the back. This homunculus is beyond consciousness, the same way that automatic, zombie-like behavior is beyond consciousness, although one is "above" it (supramental) and the other is "below" it (submental). Consciousness is an intermediate level, which is conscious not of the homunculus and its work (its "thoughts") but only of representations of the homunculus' work in the form of inner speech.

Consciousness resides at an intermediate level, which is conscious not of the homunculus and its work (its "thoughts") but only of representations of the homunculus' work in the form of inner speech. We are not conscious of the homunculus that is making decisions for us, and we are not conscious of the real world. We are only conscious of the reality that we manufacture.

Time-Based Binding Abstracted

The US biologist Charles Gray then hypothesized ("Synchronous Oscillations in Neuronal Systems", 1994) that the memory of something is generated by a stream of oscillating networks. Separate brain regions (corresponding to different categories of features) send out nervous impulses at the same frequency and the perception of an object is created by the superimposed oscillation. The brain uses frequency as a means to integrate separate parts of a perception. In this way the limited capacity of the brain can handle the overwhelming amount of objects that the world contains (the number of objects we see in a lifetime exceeds the number of neurons in the brain that would be needed to store them as images).

This theory is compatible with both Damasio's and Edelman's theories, as they all posit some type of "synchrony" for consciousness of something to emerge. According to these theories, it is time, not space, that binds.

Time-based binding almost marks a revival of "gestalt" psychology (the oscillation is, for practical purposes, a gestalt).

The Italian psychiatrist Giulio Tononi, however, noticed that one would therefore expect brain wave synchronization to increase with degrees of consciousness. Instead, when people lose consciousness, brain waves become more (not less) synchronized. Synchronized neural behavior reduces the number of possible states in which the brain can be. From the point of view of Claude Shannon's Information Theory, synchronized behavior does not "add" but "reduces" the amount of information. Tononi believes that a better model for consciousness should focus on integrated information as obtained by the vast network of the brain ("Consciousness as integrated information", 2008).

An Independent Brain

The Colombian neurophysiologist Rodolfo Llinas has interpreted these findings as a scanning system that sweeps across all regions of the brain every 25 milliseconds (40 times a second). The region of the brain containing the information about a sensation constitutes the "context" of an instance of conscious experience. The 40Hz oscillation provides the "binding" of such content into a unified cognitive act.

This wave of nerve pulses is sent out from the thalamus and triggers all the synchronized cells in the cerebral cortex that are recording sensory information. The cells then fire a coherent wave of messages back to the thalamus. Only cortex cells that are active at that moment respond to the request from the thalamus. Consciousness originates from this loop between thalamus and cortex, from the constant interaction between them. Consciousness is generated by the dialogue (or "resonating activity") between thalamus and cortex.

Consciousness is simply a particular case of the way the brain works. Other brain regions have their own temporal binding code. The motor system, for example, works at 10 cycles per second (which means that movements only occur ten times a second, not continuously). Every function is controlled by a rhythmic system that occurs automatically, regardless of what is happening to the body. Consciousness happens to be the phenomenon generated by that specific rhythmic system that operates on the brain itself.

Besides the 40-cycle-per-second, the brain has a number of natural oscillatory states: at 2 cycles per second it is sleeping. One of the brain's functions is to create images: at 2 cycles per second it creates dreams; at 40 cycles per second it creates images that represent the outside world as perceived by the senses.

In other words, the brain is always working independently of what is happening outside: during sleep, i.e. in the absence of sensorial data, that work is called "dreaming"; during the day, in the presence of sensorial data, it is called thought. The difference is that the brain's automatic dreaming is conditioned by the senses: when the senses are bombarded by external stimuli, the brain can generate only some types of thought, just like the body can generate only some types of movement. At every instant, the brain is dealing with both reality and fantasy." A person's waking life is a dream modulated by the senses".

The US neuroscientist Paul Churchland provided a detailed description of how the brain perceives sensory input (in particular vision) through what he calls "vector coding". He claims that consciousness must be based on a "recurrent" network, and Koch's 40 Hz oscillation in the cortex is a convenient candidate for a brain-wide recurrent network. That brain-wide recurrent network would be able to unify the distinct senses in one consciousness.

In this sense, therefore, consciousness does not require language, and non-linguistic animals can be conscious too. Consciousness is biological, not social (its contents may be social, such as language).

Spatiotemporal Binding

In general, dynamic processes can be coherent in one frequency band while being incoherent in another band. Resonance is the capacity of a system to respond selectively to stimuli in narrow frequency bands. Television sets and radio sets allow the viewer/listener to pick the resonant frequency (i.e. the broadcast). Once that is done, the tv set responds only to inputs (i.e. shows only the broadcasts) at that frequency. The tv set becomes part of a broader network at that frequency, while remaining totally isolated at other frequencies.

The US neurophysiologist Paul Nunez thinks that something similar takes place among neural groups: some can be bound by resonance at some frequency while, at the same time, operating independently at other frequencies. This means, in particular, that two cortical groups can be highly coherent (i.e. strongly correlated) at one frequency while being weakly coherent (i.e. weakly correlated) at another frequency.

A general property of oscillators is that weakly coupled oscillators can interact strongly when they produce appropriate resonant frequencies. The US biophysicist Bill Baird had already shown how oscillatory neural networks can account for pattern formation and recognition ("Nonlinear dynamics of pattern formation and pattern recognition in the rabbit olfactory bulb", 1986). The synchronization of oscillating networks was then the object of the research by the US mathematicians Frank Hoppensteadt and Eugene Izhikevich ("Synaptic organizations and dynamical properties of weakly connected neural oscillators", 1996).

Nunez synthesizes these studies as showing that neural groups can use rhythmic activity to communicate selectively even to groups to which they are not directly connected. It means that neural groups that are not adjacent can collaborate in performing a mental task.

Brain regions are functionally isolated at some frequencies while being functionally integrated at some other frequencies. Hence the "binding" that yields consciousness takes place both in space and in time. Consciousness is due to both local and nonlocal connections.

In general, complex adaptive systems (such as the brain) can operate over a broad range of isolation or integration. At one extreme there is functional isolation of the subsystems; at the other extreme there is global coherence. The brain, in particular, is both a set of subsystems with minimal communication (when it operates in "functional isolation" mode) and a whole that is globally coherent. Neurotransmitters can move the brain along that continuum, from granular assembly of functionally isolated subsystems to highly integrated globally coherent whole. In fact, Nunez thinks that a balanced mental state is a balanced compromise between the two extremes; and that mental diseases can be explained as overcorrelated or undercorrelated states.

Nunez notes that when one undergoes anesthesia (i.e. when "thinking" is vastly reduced), local differences tend to disappear. His interpretation is that the transition from intense mental activity to minimal mental activity corresponds to a transition from local to global neural activity.

Reverberating Loops

What Edelman and Llinas have in common is the belief that higher mental functions originates from a process of loops that reverberate through the brain (in particular, between the thalamus and the cortex, the thalamus being the source of so many crucial signals and the cortex being the newer, more sophisticated part of the brain). Their theories differ in the specific mechanism that they use but they both focus on the fact that regions of the brain are connected in a bidirectional way and that they "resonate" in response to each other, they are somehow in synch.

There are other models that exploit the same paradigm.

The Chilean neurologist Francisco Varela has claimed that there is a primary consciousness common to all vertebrates. This primary consciousness is not self-awareness but merely experience of a unitary mental state. Varela thinks that it is due to a process of "phase locking": brain regions resonate, their neurons firing in synchrony, and create a cell assembly that integrates many different neural events (perceptions, emotions, memory, etc). This coherent oscillation of neurons is primary consciousness.

The US physicist Erich Harth tried to explain consciousness by means of a process that relies on "positive" feedback. Feedback can be negative or positive. Negative feedback is the familiar one, which has to do with stabilizing a process, in particular the input with the output of the process (e.g., thermostats and car engines). Positive feedback works in the opposite direction, at the edge of instability: the signal is amplified by itself, weakening the relationship between input and output. Harth thinks that a loop of positive feedback spreads through different areas of the brain and provides "selective amplification". The loop basically joins the thalamus and the cortex, so that both send outputs that are inputs to the other. When input from the thalamus is stronger, the external world prevails. When input from the cortex is dominant, cognition prevails.

Concentric Consciousness

The British neuroscientist Susan Greenfield derived a definition of consciousness from her studies of mental infirmities: consciousness is the process of propagating a stimulus through a network of connected neurons, the same way one perceives the widening ripples created by an object falling in a pond.

She believes that, while there is no integrator of consciousness in the brain, nonetheless some type of "temporary" localization must exist. There is no site of consciousness, but consciousness originates from processes happening locally somewhere sometime. She believes that different groups of neurons take over at different times, a picture that resembles theories by Michael Gazzaniga and Daniel Dennett. Therefore, "consciousness" is multiple in space but unitary in time.

These groups of neurons she calls "gestalts": highly dynamic and transient, they are created by some kind of "arousal" and they are localized around an "epicenter".

She calls them "gestalts" because the dynamic properties of the brain "emerge" in a "gestalt" fashion from the connectivity of such neuronal assemblies.

An arousal can be an external (sensory) phenomenon or an internal (cognitive) phenomenon. The arousal causes the formation of a "gestalt" around an epicenter. The gestalt causes the "emergence" of a conscious event.

Each instance of consciousness arises from such a gestalt, caused by an arousal and localized around an epicenter. Overall consciousness develops from epicenters spread around the brain.

The passage from one conscious event to another conscious event, from one gestalt to another gestalt, which is typical of our inner life, is due to a ripple effect: the ripples of one gestalt's concentric action may act as an arousal and trigger another gestalt.

The size of a gestalt depends, first of all, on the strength of the arousal. But it also depends on the power of the epicenter to recruit neurons, a power that in turn depends on rival gestalts that all compete for neurons. The size of the gestalt has a direct meaning for us, because it corresponds to the depth of consciousness, to the intensity of the feeling.

During growth, epicenters tend to shift from "outside" to "inside", from external stimuli to internal associations.

Mental life is a dual process of searching for information and adaptation to information, the former leading to more conscious access, the latter reducing conscious access (things become habitual and automatic).

Consciousness "grows" as the brain does (from fetus to neonate to child to adult).

The Paradox of Attention

Attention is the process by which we focus on something. This sounds like an oxymoron, because the brain is always reacting to everything. Therefore it can't focus on "something". The sheer amount of data that enters the brain at every second is a distraction.

The brain is always processing everything that comes in, whether from the eyes or from the ears. The brain behaves like a machine reacting to all the inputs it receives. It doesn't have a choice: those inputs are transformed into electrochemical messages that cause electrochemical reactions spreading from the senses to the inner units of the brain. Limiting the brain's work to just one of these electrochemical processes is impossible. All the processes are going on at the same time, just because data hit our senses, and just because the senses always send signals to the brain, and just because every signal causes processing in the brain.

That is why, when we are driving and looking for a street in an unfamiliar neighborhood, we often turn down the radio. We cannot control the processing of the brain but we can control what data enter the brain.

Attention is a contradiction in terms.

The Cartesian Theater
The Dutch psycholinguist Bernard Baars believes that conscious experience is distributed widely throughout the nervous system and ultimately originates from a Darwinian process of selection applied to experience.

Consciousness creates access to a large amount of knowledge, or, more precisely, to a large number of unconscious sources of knowledge. When I am conscious of typing words into a computer, I am not aware of the keyboard, of the fact that computers are machines, of the movement of my fingers and of many other things, although each of those is essential to perform this task. All that knowledge exists, but is unconscious. Consciousness is very limited in how many things it can do (be conscious of) and for how long (it can keep them in short-term memory). But the brain as a whole does not have those limitations. Consciousness is a "gateway" to the vast knowledge stored in the brain.

In order to explain how this works, Baars employs the metaphor of a "theater".

Baars views the nervous system as a set of independent intelligent agents that broadcast messages to each other through a common workspace (just as if they were writing on a blackboard visible to every other agent). That workspace, the stage of the theater, is consciousness. Any conscious experience emerges from cooperation and competition between the many processing units of the brain working in parallel. The mind originates from the work of many independent, specialized "processors", i.e. skills that have become highly practiced, automatic and unconscious.

There are two sets of unconscious processes: the ones backstage, that, like the playwright and the director, determine what is played on the stage, and the ones in the audience, that capture what is being played. Whenever consciousness of something is created, the "audience" retrieves knowledge about that something. Consciousness is the gateway to the unconscious processes of the brain just like the stage is the gateway to the audience. Through consciousness, the backstage processes broadcast a message to all the spectator processes.

Baars emphasizes the striking differences between conscious and unconscious processes: unconscious processes are much more effective (e.g., we parse sentences unconsciously all the time, but cannot consciously describe how we parse them), they operate in parallel (whereas we can only have one conscious process at the time), they appear to have almost unlimited capacity (conscious processes have very limited capacity).

The conscious "stage" interacts with several unconscious "experts" that create goals and plans and compete for playing them on the stage. These experts are "modules" (eyesight, fear, hunger, etc) that compete for access to the "global workspace". A form of "natural selection" decides which module (or modules) is predominant at any time.

Consciousness is the stage, and the modules are the actors. Actors are competing for the spotlight.

The Social Brain
The US neurophysiologist Michael Gazzaniga, a disciple of Roger Sperry (the psychologist commonly associated with the discovery of the "split brain") showed that the brain has a modular organization, whereby many independent systems work in parallel.

These "specialized" modules are evolutionary additions to the brain: it is no surprise that they perform different functions, and it is not surprising that those functions are useful to the overall functioning of the brain.

Split-brain surgery had already proved that the human brain is made of at least two brains (the two hemispheres). Gazzaniga simply extended that idea.

Basically, he thinks that many minds coexist in a confederation. Our behavior is due to the activity of these modules, rather than to conscious decisions.

A special module, the "interpreter", located in the left hemisphere, interprets the actions of modules and provides explanations for our behavior. However, that happens after the unconscious modules have already determined our behavior. Beliefs do not precede behavior: they follow it. They are created by the interaction of the interpreter with the other modules. Homo Sapiens is the only "story-telling" animal: our brain is designed to make sense of what happens, and, in particular, of what we do.

Behavior determines our beliefs, not the other way around. It is only by behaving that we conceptualize our selves, that we build a theory of our psychological state (in particular, beliefs).

Ultimately, mental life (the life of the interpreter) is the reconstruction of the independent operations of many brain systems.

There are many "i"'s and then one "i" that makes sense of what all other "i"'s are doing.

Initially, Gazzaniga took issue with the view of the "specialized" hemispheres. It is not true that the two hemispheres are highly specialized units, and that language is "localized" in the left hemisphere: the only thing that is "lateralized" is language, and language uses up space within an hemisphere at the expense of non-linguistic features of that hemisphere.

Lateralization (the fact that each hemisphere is endowed with capabilities that the other hemisphere lacks) is not specialization.

Gazzaniga and his associates also discovered that an experience has multiple aspects which are stored at different places in the brain. These memory locations are coherent but do not seem to communicate with each other.

The only way for the brain to realize its whole knowledge is for it to watch itself as it behaves. Therefore, behavior must be the source of communication between modules. Once a memory system causes behavior, the other memory systems become "aware" of those impulses/knowledge coming from the memory system that caused the behavior to occur.

Then Gazzaniga reached another important conclusion: the conscious self is not aware of our actions before we perform them but only afterwards, although it "thinks" that it caused the action. Inspired by the 1956 theory of "cognitive dissonance" by the US psychologist Leon Festinger, Gazzaniga concludes that one of the selves, the verbal self, keeps track of what the person is doing and from that, interprets reality. As Festinger originally put it, the sense of reality arises as a consequence of considering what one does. In a sense, we become aware of our actions only "after the fact", as already discovered in 1965 by the US physiologist Benjamin Libet ("Cortical Activation in Conscious and Unconscious Experience", 1965).

Gazzaniga postulates the existence of multiple mental systems in the brain, each with the capacity to produce behavior. During growth one mental system, the verbal one, grows to "oversee" the others. The mind is not a psychological entity but a sociological entity. Language allows us to create a personal sense of conscious reality out of multiple mental realities.

The job of the verbal system is to make sense of mental activity and therefore provide the illusion of the self, when in reality there exists a conglomerate of selves.

There is still a puzzle, though, in Gazzaniga's theory: if different brain regions are perfectly capable of providing good reactions to the various situations we encounter, why is there any need for consciousness at all to make sense of the world? Why is there a need for an "interpreter"? (And one that, according to Gazzaniga's findings, makes all sorts of mistakes). If consciousness has no effect on the working of these brain regions, then it's difficult to imagine why it would appear over the course of evolution. If it does have an effect, then Gazzaniga cannot claim that it is merely an "interpreter"...

The Multimind

Reacting to the identity theory, the US psychologist Robert Ornstein developed his "multi-aspect" theory of the mind: the mental has an experiential (the experience of feeling a feeling), a neural (the corresponding brain processes), a behavioral (the related action) and a verbal (the related utterance) aspect.

The human mind is viewed by Ornstein as many small minds, each operating independently and specialized in one task. In other words, the body contains many centers of control. The lower level ones developed millions of years ago for basic survival activities, and humans share them with other animals. The most recent ones (e.g., the cortex) deal with decisions, language, reasoning. The brain is not a single whole, it is a confederation of more or less independent brains.

The goal of the mind is to simplify, to reduce the complexity of the external world to what is useful for the body. Minds are, therefore, attracted only to four types of events: recent events, unusual events, relevant events, events that can be compared to other events. When information is meaningful, it gets organized (i.e. simplified). That is the role played by the mind for the benefit of the body.

The mind is an adaptive system that has been shaped by the world. It is the way it is because the world is the way it is.

Human minds are initially endowed with many possible ways of evolving (e.g., with the capability for learning many possible languages), but only some are pursued and the other skills are lost during growth. The mind could potentially adapt to many different environments, but will actually adapt only to the ones it is exposed to. During development, a number of specialized, autonomous centers of action develop. These "minds" within the mind compete for control of the organism. Each one tends to stay in place for as long as possible, with its own memories and goals, until the circumstances favor the takeover of another mind.

In a sense, this is a Darwinian model, very similar to Edelman's model for the neural structure of the brain. There are many selves that take control of the brain and succeed or fail based on natural selection. Some die out, and some get stronger, depending on how often they succeed.

Multiple Drafts

The US philosopher Daniel Dennett believes that, despite the apparent unity and continuity of our experience, consciousness is non-localized and non-linear. "Non-localized" means that there is no place where it happens. "Non-linear" means that it is not a flow of feelings.

Dennett argues that the time-scale of some cognitive processes rules out the possibility that perceptions are integrated in a "Cartesian theater" in the brain before action is generated (the Cartesian theater is a metaphor for the idea that there is a central locus in the brain that directs consciousness). Despite the apparent unity of our experience, consciousness does not involve the existence of a single central self.

Dennett, instead, argues that the mind is occupied by several parallel "drafts". A "draft" can be roughly viewed as a narrative that occurs in the mind, and that is typically triggered by some interaction with the environment. At every point in time, one of those narratives is dominant in the brain, and that is what we are conscious of. "Consciousness" is a vague term, which simply refers to the feeling of the overall brain activity. But the truth is that there are many drafts, all working in parallel. There are several narratives in the mind going on at the same time.

Dennett is opposed to the idea that there is an enduring mind because it would imply that there is a place in the brain where that mind resides. He thinks that such "Cartesian theater" is absurd and that the mind is implemented by multiple parallel drafts.

A mental content becomes conscious by winning the competition with other mental contents and therefore lasting longer in the mind. A mind is an organization of competing mental events.

Despite the apparent continuity of our experience, consciousness does not flow at all. There is no single stream of consciousness à la William James because there is no "Cartesian theater", no central control. There are parallel circuits, which produce parallel drafts of narratives. The continuity of consciousness is an illusion.

Consciousness doesn't even exist all the time, as "probing precipitates narratives": people are not always conscious of what is happening to them until, for example, somebody else asks them about it. We become conscious when something unexpected happens: as Karl Popper put it, we don't "hear" (pay attention to) the clock ticking, but we "hear" that it stopped ticking.

Dennett also believes that the goal of those drafts, the goal of the mind, is truly to manage "memes". The mind was created, evolutionary speaking, when memes "invaded" the brain. The mind of each individual is created little by little as memes invade it. The brain has become a computer that collects memes. The mind is a machine to process memes, not too different from the body, the machine that processes genes. Consciousness is but a collection of memes.

The British psychologist Susan Blackmore has expressed the same view in more radical words. The conscious self is but a story built by memes. In this sense, it makes no sense to talk of free will. Free will is the consequence of the story (the very complex story due to many interacting memes) that is playing in the brain.

Dennett views religions and ideologies as memes that spread from mind to mind. These memes created our consciousness by degrees (mildly put, they had a tremendous influence on our thinking). Memes such as religions were originally ideas that sprung into somebody's mind. Jesus' meme, for example, was that all humans are alike: you are like me, regardless of your name, tribe or country. Each meme like this creates a new degree of consciousness. Around the world religions and ideologies made people think and become aware of the others, of death, of the world. They are contagious: once somebody talks to you about everlasting life, you can't stop wondering about it and will tell someone else. Each religion introduced minds to novel ideas about the world. Before the birth of science and ideology, religion was the main way in which minds kept expanding their consciousness.

Ultimately, thinking is processing of memes: our "mind" is the process of absorbing, understanding, adapting and broadcasting memes.

People tell us what to think and we think their thoughts. What we think are other people's thoughts. Are any of our thoughts "ours"? Or are we only vehicles for thoughts to spread from mind to mind?

However, one could argue that the "reproduction" of a meme is often not conscious (e.g., when I start whistling a song that I heard on the radio). After all, i am truly conscious only when I rid myself of memes. Then I can focus on what is truly "me". Consciousness is a failure of memes: the more powerful the memes the less conscious your mind. And, conversely, the more conscious your mind the less powerful the memes that try to invade it.

Mental Darwinism

These ideas all point in the same direction, towards a Darwinian picture of consciousness.

The US neurophysiologist William Calvin has put forth a theory called "mental Darwinism" that neatly abstracts this view.

Just like the immune system and the evolution of species are driven by natural selection, mental life too is driven by natural selection. A Darwinian process in the brain finds the best thought from the many that are produced continuously. Thoughts evolve subconsciously. As Carl Jung said, dreaming occurs all the time but we can't see them when we are awake.

Calvin's emphasis is on movement: ultimately, what we think is for the sake of action. We need to move in the world and the mind is one tool to determine the most efficient way to do that. Far from being merely combinations of sensations and memories, thoughts are movements that haven't happened yet.

To explain how this works, Calvin has introduced the concept of a "cerebral code". Cerebral code is the equivalent of the genetic code: it allows for reproduction and selection of thoughts. A cerebral code copies itself repeatedly around a region of the brain, in a manner similar to Donald Hebb's cell assemblies. Thought arises from the copying and competition of cerebral codes. Our actual thought is simply the dominant pattern in the copying competition. The brain is an evolutionary system. The brain is what Calvin calls a "Darwin Machine".

He has shown that circuits in the cerebral cortex act as copying machines, but they copy in a Darwinian fashion, introducing mistakes that continuously create variants. Such variants then compete for cortex space. Calvin claims to have found in the brain all Darwinian algorithms for evolution, and even the catalysts that speed up evolution, i.e. the equivalents of sex, island settings and climate change.

Uncentered Consciousness

Based on the fact that brain lesions remove consciousness only when they remove performance, the Austrian psychologist Marcel Kinsbourne ("Integrated Field Theory of Consciousness", 1988) reached the conclusion that consciousness "is" performance, and developed an "integrated cortical field theory" of consciousness.

Kinsbourne criticizes the idea that consciousness sits at the top of a pyramid of cognitive faculties and that it is produced by the neural activity of a specialized region of the brain.

Kinsbourne believes that consciousness is not a product of neural activity: it is the neural activity itself. The brain does not generate consciousness: it is conscious. There is no need for a special region to manufacture consciousness. Kinsbourne criticizes the idea that for some information to become conscious it has to be input to a special region of the brain, which is in charge of "transducing" neural activity into a conscious feeling. Kinsbourne believes that it is not the region that matters but the state of the circuit. Any region of the brain can be conscious when its circuits are in the appropriate state.

There is no central site for consciousness: any site can host consciousness. In this model there is no need for binding. And the model admits the possibility of more than one consciousness. Kinsbourne's model is "heterarchical", i.e. highly distributed.

The US neuroscientist Denise Ingebo-Barth believes that consciousness is a stream of discrete conscious events, just like a film is made by a stream of frames. The nervous system operates based on its own programming and its input. An operation of the nervous system results in a "trajectory" of neural events in the brain. Whenever that trajectory crosses the thalamus, a conscious event is generated. That trajectory could be coming from the senses (and result in a sensation) or the cortex (and result in a thought) and could be going to the cortex (and result in a memory). Trajectories evolve through a series of possible "fluctuations" and superimpositions, and, when they cross the thalamus, result in feelings.

The Dream of Consciousness
Scientists who studied dreams, such as Jonathan Winson and Allan Hobson, ended up believing that dreams hold the secret to consciousness, and that consciousness may simply be a consequence or manifestation of the same process that creates dreams.

Winson believes that the "subconscious" is an ancient mechanism involving REM sleep, according to which memories and strategies were formed. Dreams were helping us survive a long time before our mind was capable of providing any help at all. The mind could then be but an evolution of dreaming. First the brain started dreaming, then dreams took over the brain and became the mind. Maybe the mind is simply one long, continuous dream of the universe.

This account of how the mind came to be is similar to the hypothesis that the mind was created by memes. Dreams and memes share the property of "invading" the mind, although one is private while the other is public (as Joseph Campbell's aphorism goes, "a myth is a public dream, a dream is a private myth").

Hobson's theory of dreams focused on identifying the two chemical systems inside the brain that regulate the waking and the dreaming experience (the "aminergic" and the "cholinergic" systems). Hobson came to believe that the interplay of these chemical processes is responsible for all of consciousness.

Conscious states fluctuate continuously between waking and dreaming. Even at the extremes, both chemical systems are active. Between the extremes there is a continuum of states which explains phenomena such as hypnosis, fantasy, concentration, etc. The three fundamental states of consciousness are waking, sleeping and dreaming. Hobson's model of AIM (Activation/ Information/ Mode) attempts to identify the quantities that regulate the transition from one state to another.

According to Hobson, mind is more than consciousness (parts of it are unconscious) and dreams are part of consciousness.

Consciousness is a graded characteristic.

"Mind" is all the information in the brain. Consciousness is the brain's awareness of some of that information.

Ultimately, consciousness is the brain's representation of the world, the body and the self. Consciousness is a representation by the brain of the representation by the brain of the world, the body and the self. Adult humans possess the brain circuitry to achieve this "representation of a representation". That circuitry constitutes a "secondary" network within the brain that is responsible for "meta-representational" functions.

Time

A lot of what "consciousness" is depends on time. But time is the ever elusive quantity. The Hungarian philosopher Julius Fraser is among those who think that we use the same word, "time", for different things. Fraser has resurrected the concept of "umwelts" (a term originally introduced by Jacob von Uexkull to refer to each species' unique experience of the world) and applied it to different levels of organization of Nature. At each level the world appears in a different light.

At the lowest level of organization, the level of pure energy, time does not exist: the lowest level is atemporal.

At the quantum level (the level studied by Quantum Physics), reality oscillates between waves and particles, which means that time oscillates between existence and non-existence: the quantum level is prototemporal. At this level causality is stochastic.

At the chemical level, there is time, and it is the time of classical Physics, the time that we can measure and that is absolute. Classical Physics can be played in either direction, so time has no preferred direction: the classical level is eotemporal. At this level causality is deterministic. There is no "now".

At the biological level, time assumes a specific direction: one can only go forward (live and die), never backwards. And "now" is a fundamental concept. The biological level is biotemporal. At this level causality is emergent.

At the conscious level, the forward direction is even stronger, because we can even experience time before we existed and after we will cease to exist. The conscious level is nootemporal.

Human time contains all of these forms of "time". Human time is a recapitulation of the evolution of consciousness through all those levels.

Thinking About Time

Conscious beings have a built-in feeling of time. Time feels continuous and feels ordered in a chronology of events.

Neither the continuity of time (all the astronomical, geological and biological "clocks" are quantized in nature) nor the ordering of time can be easily explained. They are both quite "unnatural".

Nature uses different clocks at different levels of organization. There are seasons (that trigger life stages such as hibernation or pollination). There is the alternation of day and night (sleep, feeding). Then there are countless body clocks that regulate everything from heartbeat to eyesight. At the lowest level of organization there are clocks for the cells. Since the feeling of time is in the brain, the neurons are particularly interesting kinds of cells. The hippocampus is a region of the brain that is crucial for the maintenance of long-term memories, and therefore to the sense of flowing time.

The "clock" of the hippocampus is represented by theta oscillations. The Greek neurologist Thanos Siapas ("Hippocampal theta oscillations are traveling waves", 2009) has shown that they are not synchronized throughout the hippocampus. Taken together, the theta oscillations constitute a traveling wave through the hippocampus. They travel at a speed of about 100 mm/sec, which means that their spatial length-wave is approximately the size of the hippocampus. In other words, they behave like the rhythm of day and night around the various time zones of the Earth. This wave of theta oscillations "paces" the hippocampus in a manner similar to a regular progression of time zones.

Biological organisms need to deal with time at at least twelve orders of magnitude, from 10 to the -6th seconds (neurons) to 10 to the 6th seconds (days). Humans use the same technological artifact (clocks) to measure time at all of these levels. The brain, however, does not seem to have a clock. There is no central timekeeper in the brain. The brain uses different mechanisms to measure time depending on time scale. There are several techniques that the brain can use to measure time. The first one is the same one used by clocks: an oscillator and a counter. The second one is to encode timing in the firing rate of neurons. A third one, propounded by the US neurologist Dean Buonomano ("State-dependent Computations"), is that our perception of time might be a consequence of the organization of neurons in networks. Timing emerges in neural networks as a result of their dynamics in the face of stimuli. The strengths of synapses change in time like ripples in a pond. Just like the extent of a ripple is an indication of how much time has elapsed since the pebble hit the water, so the dynamics of synapses can be used to "measure" elapsed time. Unlike clocks that "tick" in a linear manner, the brain does not have a ticking clock and therefore does not keep track of time in a linear manner.

The Deception of Consciousness

The Danish mathematician Tor Norretranders pointed out that the mind is more than what we feel.

The senses process huge amounts of information, but consciousness contains almost no information at all. Most "mental" life never becomes conscious: it is lost in the processing. Large quantities of information are discarded before consciousness occurs. The discarded information, nonetheless, has an influence on our behavior. There is a non-conscious aspect of the human experience that we are not familiar with because we cannot "feel" it.

But this also means that consciousness is mostly about what happens inside us, not what happens outside. Sense data are processed according to our brain structure and matched with data in memory, and processed again, and then a conscious feeling arises. Very little of the original sense data is present when the feeling arises.

Sense data are filtered by countless neural processes in the brain before they become conscious sensations: we cannot experience the sense data, the original. We can only experience the finished product, never the raw material. We only experience a bit of what our body experiences and even that "bit" is not exactly what the body experienced but a "doctored" version of it.

The paradox is that our brain knows more than our consciousness does.

There is self-deception on the part of consciousness: before we experience it, the content of consciousness has been processed and transformed from its original format. Consciousness presents us with an altered, subjective, tampered with view of reality but doesn't tell us so.

Norretranders separates the conscious (thinking) "i" from the acting (instinctive) "me". The "i" is held responsible for the actions of the "me", although the "i" is often not aware (literally) of what the "me" is doing.

Consciousness is a Place

Consciousness is a place. I don't feel myself in a foot or a hand: I feel myself in the head. I don't feel myself outside my body. If you built an exact replica of my body, I would still feel myself only inside my original body (and "someone else" would feel himself into the replica of my body).

The mother of all problems is that the human brain is not the most intuitive structure to generate consciousness. If we asked a team of engineers (who have never seen a lake or a mountain) to build a reservoir of water, they might come up with the same alpine lake that nature created. If we asked them to create an organism that can survive at high elevation, they might design a marmot. But even if you put the smartest engineers in the world in a room and asked them to design something that is conscious, nobody would come up with an oddly shaped jelly of bilions of interconnected neurons. The brain is such an unlikely structure (and, let's face it, a fairly ugly one).

Consciousness and Complexity

It has become fashionable to explain how consciousness emerges from the brain by invoking the emergent properties of complex adaptive systems. The problem is that the brain is not the only complex adaptive system inside the human body. The eye, the heart, the stomach and pretty much every other organ are complex and adaptive too, but they don't seem to have much to do with consciousness, not even a faint one. The property of complexity alone does not explain consciousness. The question is what makes the complex system of the brain different from the complex system of any other bodily organ.

The US neurophysiologist Paul Nunez emphasizes the cortex's nested hierarchical structure ("neurons within minicolumns within modules within macroculumns inside cerebral cortex lobes"), a model pioneered in the 1970s by the US neurobiologist Walter Freeman.

That might be the unique feature of the brain, but, to some extent, many other bodily organs and natural systems exhibit a "nested hierarchical" organization: are they conscious too? For example, any muscle is a complex adaptive system that learns from experience, just like the brain. It seems odd that consciousness is a zero or one: the highly nested hierarchical complexity of the brain yields my consciousness, while the less nested hierarchical complexity of the heart yields absolutely no consciousness. It would be more logical to find a continuum of contributions to my consciousness by all the complex organs of my body all the way down to fingertips and hairs. Consciousness should be distributed all over the body, to different degrees.

Of course, an alternative theory is that every complex system, including my eyes and my heart, generate some kind of conscious life, and the one that is writing these words, that can communicate with other consciousnesses, and that can teach cognitive science in universities is the one created by the brain. Hence the literature (created by brain-generated consciousnesses) will only know of brain-generated consciousnesses and never know of, say, heart-generated consciousnesses.

Between Conscious and Subconsciouss

Everybody agrees that there are conscious states of mind and unconscious states of mind. Most of what we do is done unconsciously. When you drive to work, you perform an extremely sophisticated task that requires driving skills, orientation skills, and avoiding all the obstacles and reckless drivers on the road; but usually you do it while listening to the radio a humming a song, hardly focusing on the action itself. On the other hand, there are moments when you pause and wonder, very consciously, about the meaning of life, or, if you are a poet, about your depressing financial conditions. However, these two extremes do not exhaust the kinds of mental life that we have. For example, what you are doing now, reading these sentences, is not quite conscious and not quite unconscious. When you read, you have to focus on the sentences. In fact, your attention increases, not decreases, compared with other moments. However, you are not quite conscious of your existence, precisely because you are so focused on what you are reading. The "reading" itself is effortless, just like many other routine tasks that you perform during the day without "thinking". But it would be unfair to all it "unconscious" since you have to push yourself to keep reading a difficult book like mine. In a sense, when you read, you are not conscious of yourself but you are as attentive as you can be. You can also feel strong emotions, like when you are reading a sentimental novel or a newspaper article about a brutal homicide.

Consciousness is usually about memorizing. When you are driving the familiar route to work, you hardly memorize anything that happens along the way, unless it is truly unusual (e.g. a car accident, a fire, a fallen tree). When you are unconscious of your actions, there is little or no memorizing going on. Their mostly reinforcement of knowledge you already had memorized. To memorize something, you have to focus on what you want to memorize, or at least you have to be devoting your full attention to that event. But when you are reading a book like this one, with the intent of memorizing as much information as possible, you are not quite conscious of what you are doing while actually at the peak of your learning activities. Studying, in fact, requires that you don't "think", that you direct your entire mental activity on the textbook; and that mental activity results in maximum memorizing. Is that activity conscious or unconscious?

Degrees of Consciousness

There is a level of consciousness that we do not achieve all the time, and some people may never or rarely achieve. Teenagers tend to watch movies and wear clothes and even eat food depending on which message prevails in marketing campaigns. The kids who do not follow the "trend" are considered "freaks", "weirdos". The truth is that the weird kids are probably more conscious than their friends. The weird kids are probably the ones who realize that there is nothing special about that actor or that fast-food chain: it is just that a lot of money has been spent to promote them.

As people grow up they tend to be more aware of why they do things. Some keep following the trends while others become more and more individualistic. The latter are invariably labeled as "eccentric": "eccentric" means that you use your brain instead of letting an external phenomenon condition your brain. The more "conscious" you are, the more eccentric they think you are.

The Beatles, Hollywood stars, Coca Cola and McDonald's are all examples of things that people like because they have been told to like them. The less conscious somebody is, the more dependent that person will become on those things.

Society is built on a careful balance of thinkers and non-thinkers. Society relies on a few thinkers to break the rules and bring innovation, but at the same time relies on non-thinkers to perform the daily tasks that keep society alive. Business relies on a few people to bring innovation and on millions of non-thinkers to buy it. Capitalism, communism, fascism all rely on people not to think too much, otherwise the system would become highly unstable.

If people thought more, McDonald's and Coca Cola would be out of business, and only a minority would listen to pop stars or watch Hollywood blockbusters. And probably very few people would work as hard as they do. It would be total anarchy.

Not only do different levels of consciousness exist, but society (and possibly nature at large) depends on a delicate balance of those levels. A successful society is made possible by the very fact that its members have different degrees of consciousness.

Further Reading
Baars, Bernard: A COGNITIVE THEORY OF CONSCIOUSNESS (Cambridge Univ Press, 1988)

Baars, Bernard: IN THE THEATER OF CONSCIOUSNESS (Oxford Univ Press, 1997)

Bateson, Gregory: MIND AND NATURE (Dutton, 1979)

Blackmore, Susan: THE MEME MACHINE (Oxford Univ Press, 1999)

Calvin, William: HOW BRAINS THINK (Basic, 1996)

Cannon, Walter: THE WISDOM OF THE BODY (Norton, 1939)

Cohen, Jonathan & Schooler, Jonathan: SCIENTIFIC APPROACHES TO CONSCIOUSNESS (Erlbaum, 1997)

Churchland, Paul: ENGINE OF REASON (MIT Press, 1995)

Damasio, Antonio: DESCARTES' ERROR (G.P. Putnam's Sons, 1995)

Damasio, Antonio: THE FEELING OF WHAT HAPPENS (Harcourt Brace, 1999)

Dawkins, Richard: THE SELFISH GENE (1976)

Dennett, Daniel: CONSCIOUSNESS EXPLAINED (Little & Brown, 1991)

Dennett, Daniel: KINDS OF MINDS (Basic, 1998)

Deutsch David: THE BEGINNING OF INFINITY (Viking, 2011)
Dretske, Fred: NATURALIZING THE MIND (MIT Press, 1995)
Eccles, John: EVOLUTION OF THE BRAIN (Routledge, 1991)
Edelman, Gerald: THE REMEMBERED PRESENT (Basic, 1989)
Edelman, Gerald: BRIGHT AIR BRILLIANT FIRE (Basic, 1992)
Edelman, Gerald & Tononi Giulio: A UNIVERSE OF CONSCIOUSNESS (Basic, 2000)
Farthing, G.William: THE PSYCHOLOGY OF CONSCIOUSNESS (Prentice-Hall, 1992)
Festinger, Leon: THEORY OF COGNITIVE DISSONANCE (1957)
Flanagan, Owen: THE SCIENCE OF THE MIND (MIT Press, 1991)
Flanagan, Owen: CONSCIOUSNESS RECONSIDERED (MIT Press, 1992)
Fraser, Julius: OF TIME, PASSION AND KNOWLEDGE (Princeton Univ Press, 1975)
Gazzaniga, Michael: SOCIAL BRAIN (Basic, 1985)
Gazzaniga, Michael: NATURE's MIND (Basic, 1992)
Gazzaniga, Michael: THE MIND's PAST (Univ of California Press, 1998)
Greenfield, Susan: THE HUMAN MIND EXPLAINED (Henry Holt & Co, 1996)
Grof, Staninslav: THE HOLOTROPIC MIND (Harper, 1993)
Haldane, John: Possible Worlds and Other Papers (1927)
Harth, Erich: CREATIVE LOOP (Addison-Wesley, 1993)
Hobson, Allan: THE CHEMISTRY OF CONSCIOUS STATES (Little & Brown, 1994)
Humphrey, Nicholas: CONSCIOUSNESS REGAINED (Oxford Univ Press, 1983)
Humphrey, Nicholas: A HISTORY OF THE MIND (Simon & Schuster, 1993)
Ingebo-Barth, Denise: THE CONSCIOUS STREAM (Universal Publisher, 2000)
Jackendoff, Ray: CONSCIOUSNESS AND THE COMPUTATIONAL MIND (MIT Press, 1987)
Jackson, Frank & Braddon-Mitchell, Basil: THE PHILOSOPHY OF MIND AND COGNITION (Basil Blackwell, 1996)
Kinsbourne, Marcel: TIME AND THE OBSERVER (1992)
Koch, Christof: THE QUEST FOR CONSCIOUSNESS (Roberts, 2004)
Llinas, Rodolfo & Churchland Patricia: THE MIND-BRAIN CONTINUUM (MIT Press, 1996)

Lycan, William: CONSCIOUSNESS AND EXPERIENCE (MIT Press, 1996)

McGinn, Colin: THE PROBLEM OF CONSCIOUSNESS (Oxford Univ Press, 1991)

McGinn, Colin: THE MYSTERIOUS FLAME (Oxford Univ Press, 1999)

Mead, George Herbert: MIND SELF AND SOCIETY (Univ of Chicago Press, 1934)

Mead, George Herbert: THE PHILOSOPHY OF THE ACT (Univ of Chicago Press, 1938)

Nagel, Thomas: MORTAL QUESTIONS (Cambridge Univ Press, 1979)

Nagel, Thomas: THE VIEW FROM NOWHERE (Oxford Univ Press, 1986)

Neisser, Ulric: CONCEPTS AND CONCEPTUAL DEVELOPMENT (Cambridge University Press, 1987)

Norretranders, Tor: THE USER ILLUSION (Viking, 1998)

Nunez, Paul: BRAIN, MIND, AND THE STRUCTURE OF REALITY (Oxford Univ Press, 2010)

Ornstein, Robert: MULTIMIND (Houghton Mifflin, 1986)

Ornstein, Robert: EVOLUTION OF CONSCIOUSNESS (Prentice Hall, 1991)

Russell, Bertrand: THE PROBLEMS OF PHILOSOPHY (1912)

Scott, Alwyn: STAIRWAY TO THE MIND (Copernicus, 1995)

Searle, John: THE REDISCOVERY OF THE MIND (MIT Press, 1992)

Tononi, Giulio: PHI, A VOYAGE FROM THE BRAIN TO THE SOUL (Knopf Doubleday, 2012)

Tye, Michael: THE METAPHYSICS OF MIND (Cambridge University Press, 1989)

Tye, Michael: TEN PROBLEMS OF CONSCIOUSNESS (MIT Press, 1995)

Unger, Peter: IDENTITY, CONSCIOUSNESS AND VALUE (Oxford Univ Press, 1991)

Varela, Francisco, Thompson, Evan, Rosch, Eleanor: THE EMBODIED MIND (MIT Press, 1995)

Winson, Jonathan: BRAIN AND PSYCHE (Anchor Press, 1985)

Zeman, Adam: CONSCIOUSNESS (Yale Univ Press, 2002)

THE PHYSICS OF CONSCIOUSNESS

A Critique of Neuroscience

Neuroscience of the 20th century was based on classical Physics. No surprise, then, that it derived a view of the brain as a set of mechanical laws: that is the "only" view that classical Physics can derive. No surprise that it could not explain how consciousness arises, since there is no consciousness in classical Physics: it was erased from the study of matter by Descartes' dualism (that mind and matter are separate), on which foundations Newton erected classical Physics (the science of matter, which does not deal with mind). By definition, Descartes' dualism predicts that "mind" cannot be explained from matter, and Newton's Physics is basically an instantiation of Descartes' dualism. Which means that Descartes' dualism predicts that Newton's Physics cannot explain the conscious mind. Neuroscientists of the 20th century who were looking for consciousness missed that simple syllogism: they were looking for consciousness using a tool that was labeled "not suitable for finding consciousness".

Neuroscience of the 20th century rested on the Newtonian principle that a physical system is made of independent parts which interact only with their immediate neighbors and whose behavior over time is deterministic. Within this paradigm, a mind is the product of a brain, which is one particular system of the many that populate the universe. This is a useful paradigm for the study of many material phenomena, but it is not what the Physics of the 20th century prescribed. It is what Physics prescribed a century earlier, before it was showed to be wrong.

Neurological descriptions of the brain that are based on Newton's Physics are based on a Physic that is known to have limitations at a small and large scale. Twentieth century neurologists assumed that the brain and its parts behave like classical objects, and that quantum effects are negligible, even while the "objects" that they were studying got smaller and smaller. What 20th century neurologists were doing when they studied the microstructure of the brain from a Newtonian perspective was equivalent to organizing a trip to the Moon on the basis of Aristotle's Physics, neglecting Newton's theory of gravitation.

Quantum Consciousness

If no theory of consciousness based on classical Physics is satisfactory in explaining how consciousness emerges from the electrochemical activity of the brain, then maybe the problem lies with classical Physics. Physicists began in the 1920s to advocate an approach to consciousness based on 20th-century Physics rather than classical Physics.

Loosely speaking, the point is that consciousness is unlikely to arise from classical properties of matter (the more we understand the structure and the electrochemical fabric of the brain, the less we understand how consciousness can occur at all). But, for example, Quantum Theory allows for a new concept of matter altogether, which may well leave cracks for consciousness, for something that is not purely material or purely extra-material.

Of course, the danger in this way of thinking is to relate consciousness and Quantum Theory only because they are both poorly understood: what they have in common is a degree of "fuzziness" that allows us to tinker with definitions.

The advantage of Quantum Theory, though, is that it allows for "non-local" properties and provides a framework to explain how entities get "entangled", precisely the phenomena that electrochemical brain processes are not enough to explain.

The unity of consciousness is a favorite example. A conscious state is the whole of the conscious state and cannot be divided into components (I can't separate the feeling of red from the feeling of the apple when I think of a red apple). Newton's Physics is less suitable than Quantum Theory for dealing with such a system, especially since Bell's Theorem proved that everything is permanently interacting. Indeterminate behavior (for example, free will) is another favorite, since werner Heisenberg's principle allows for some unpredictability in nature that Newton's Physics ruled out. And, of course, the mind/body dualism reminds Physicists of the wave/particle dualism. In fact, René Descartes' dualism is less credible within the framework of Quantum Physics because, in Quantum Physics, matter is ultimately not a solid substance.

Quantizing the Mind

The pioneer of "quantum consciousness" theories was the Ukrainian chemist Alfred Lotka, who in 1924, when Quantum Theory was still in its infancy, proposed that the mind controls the brain by modulating the quantum jumps that would otherwise lead to a completely random existence.

The first detailed quantum model of consciousness was probably the US physicist Evan Walker's synaptic tunneling model ("The Nature of Consciousness", 1970), in which electrons can "tunnel" between adjacent neurons, thereby creating a virtual neural network overlapping the real one. It is, Walker claims, this virtual nervous system that, according to

Walker, produces consciousness and that can direct the behavior of the real nervous system.

Walker views consciousness as a different domain of discourse from matter, but, at the same time, recognizes that it is affected by matter: therefore, there must be a way that consciousness and the material world can interact. In particular, the nature of consciousness must be such that it is directly related to events in the brain. Walker based his theory on two postulates: 1. Consciousness is real and nonphysical; and 2. Physical reality is connected to consciousness by a physically fundamental quantity. Walker believes that the quantum tunneling effect satisfies both postulates. He can even write the equation for consciousness (the number of electrons that, thanks to the tunneling effect, manage to connect two active synapses). As a result, the observer of Quantum Physics turns out to be a quantum system herself.

Consciousness is the set of potentialities created by the tunneling effect across the brain. But only a portion of that set becomes reality, as only some potentialities are realized when the wave function collapses at the synapsis. Walker uses that subset to define another mental quantity: "will". Our will is distinct from our consciousness in that our consciousness contains all the possibilities, whereas our will is only what actually happens (what the body actually does).

Following the Hungarian physicist Eugene Wigner, Walker proposes to add a term to Schroedinger's equation that would make it nonlinear and that would explain what causes the collapse of the wave: a measurement of information. This term, that basically expresses the transfer of information that takes place with the wave's collapse, would disappear once the measurement is performed. Basically, this term would signal the presence of the observer. By introducing the same "information term" in Paul Dirac's equation, Walker derives another possible interpretation: reality is consciousness observing itself. Dirac's equation becomes simply the equation of an observer observing.

The "real" nervous system operates by means of synaptic messages. The virtual one operates by means of the quantum effect of tunneling (particles passing through an energy barrier that classically they should not be able to climb). The real one is driven by classical laws; the virtual one by quantum laws. Consciousness is, therefore, driven by quantum laws, even though the brain's behavior can be described by classical laws.

Walker interprets Albert Einstein's four-dimensional space-time as time plus an ordering of events that were probable but did not happen (something that we call "space"). The only thing that exists, ultimately, is the observer, who consciously experiences her complement. The sequence of conscious experiences is time, and the set of possible events is space. The universe is the observer observing.

Later theories share with Walker's the view that the brain "instantiates" not one but two systems: a classical one and a quantum one; the second one being responsible for the properties of mental life (such as consciousness) that are not easily reduced to the properties of the classical brain.

The British neurologist John Eccles speculated that synapses in the cortex respond in a probabilistic manner to neural excitation ("Do Mental Events Cause Neural Events Analogously To The Probability Fields Of Quantum Mechanics?", 1986). That probability might well be governed by quantum uncertainty given the extremely small size of the synapsis' microscopic organ that emits the neurotransmitter. Eccles speculates that an immaterial mind (in the form of "psychons") controls the quantum "jumps" and turns them into voluntary excitations of the neurons that account for body motion.

Drawing from Quantum Mechanics and from Bertrand Russell's idea that consciousness provides a kind of "window" onto the brain, the philosopher Michael Lockwood advanced a theory of consciousness as a process of perception of brain states. First he noted that Special Relativity implies that mental states must be physical states (mental states must be in space given that they are in time). Then Lockwood interpreted the role of the observer in Quantum Mechanics as the role of consciousness in the physical world (as opposed to a simple interference with the system being observed). Lockwood argued that sensations must be intrinsic attributes of physical states of the brain: in quantum lingo, each observable attribute (e.g., each sensation) corresponds to an observable of the brain. Consciousness scans the brain to look for sensations. It does not create them: it just seeks them.

There are also models of consciousness that invoke other dimensions. The unification theories that attempt at unifying General Relativity (i.e. gravitation) and Quantum Theory (i.e., the weak, electrical and strong forces) typically add new dimensions to the four ones we experience. These dimensions differ from space in that they are bound (actually, rolled up in tiny tubes) and in that they only exist for changes to occur in particle properties. The hyperspace of the US physicist Saul-Paul Sirag, for example ("Consciousness - A Hyperspace View", 1993), contains many physical dimensions and many "mental" dimensions (time is one of the dimensions that they have in common).

Bose-Einstein Condensates

Possibly the most popular candidate to yield quantum consciousness has been Bose-Einstein condensation (theoretically predicted in 1925 and first achieved in a gas in 1995). The most popular example of Bose-Einstein condensation is superconductivity.

The fascination with Bose-Einstein condensates is that they are the most highly ordered structures in nature (before their discovery in 1925 by Albert Einstein and Satyendranath Bose, that record was owned by crystals). The order is such that each of their constituents appears to occupy all their space and all their time: for all purposes the constituents of a Bose-Einstein condensate share the same identity. In other words, the constituents behave just like one constituent (the photons of a laser beam behave just like one photon) and the Bose-Einstein condensate behaves like one single particle. Another odd feature of Bose-Einstein condensates is that they seem to possess a primitive form of free will.

A Bose-Einstein condensate is the equivalent of a laser, except that it is the atoms, rather than the photons, that behave identically, as if they were a single atom. Technically speaking, as temperature drops, each atom's wave grows, until the waves of all the atoms begin to overlap and eventually merge. After they have merged, the atoms are located within the same region in space, they travel at the same speed, they vibrate at the same frequency, etc.: they become indistinguishable. The atoms have reached the lowest possible energy, but Heisenberg's principle makes it impossible for this to be zero energy: it is called "zero-point" energy, the minimum energy an atom can have.

The first Bose-Einstein condensate was created in 1995 by the US physicists Eric Cornell and Carl Wieman. In 2003 both the Austrian physicist Rudolf Grimm and the US physicist Deborah Jin achieved a Bose-Einstein "super molecule", i.e. a collection of molecules (not just atoms) behaving in perfect unison.

The intriguing feature of a Bose-Einstein condensate is that the many parts of a system not only behave as a whole, they become a "whole". Their identities merge in such a way that they lose their individuality.

It was thought that Bose-Einstein condensation could be achieved only at very low temperatures. The British physicist Herbert Froehlich ("Long-range coherence and energy storage in biological systems", 1968) proved the feasibility, and even the likelihood, of Bose-Einstein condensation at body temperatures in living matter (precisely, in cell membranes). This opened the doors to the possibility that all living systems contain Bose-Einstein condensates.

He argued that electrically charged molecules of living tissues behave like electric dipoles. When digestion of food generates enough energy, all molecular dipoles line up and oscillate in a perfectly coordinate manner, which may result in a Bose-Einstein condensate.

Biological oscillators of this kind are pervasive in nature: living matter is made of water and other biomolecules equipped with electrical dipoles, which react to external stimuli with a spontaneous breakdown of their rotational symmetry.

The biological usefulness of such biological oscillators is that, like laser light, they can amplify signals and encode information (e.g., they can "remember" an external stimulus).

Above all, coherent oscillations are crucial to many processes of integration of information in the brain.

Quantum Self Theory

The British psychiatrist Ian Marshall ("Consciousness and Bose-Einstein condensates", 1989) showed similarities between the holistic properties of condensates and those of consciousness, and suggested that consciousness may arise from the "excitation" of such a Bose-Einstein condensate. In Marshall's hypothesis, the brain contains a Froehlich-style condensate, and, whenever the condensate is excited by an electrical field, conscious experience occurs. The brain maintains dynamical coherence (i.e., the ability to organize millions of neural processes into the coherent whole of thought) thanks to an underlying quantum coherent state (the Bose-Einstein condensate).

Furthermore, Marshall thinks that the collapse of a wave function is not completely random, as predicted by Quantum Theory, but exhibits a preference for "phase difference". Such "phase differences" are the sharpest in Bose-Einstein condensates. This implies that the wave function tends to collapse towards Bose-Einstein condensates, i.e. that there is a universal tendency towards creating the living and thinking structures that populate our planet. Marshall views this as an evolutionary principle inherent in our universe.

In other words, the universe has an innate tendency towards life and consciousness. They are ultimately due to the mathematical properties of the quantum wave function, which favors the evolution of life and consciousness.

Marshall thinks we "must" exist and think, in accordance with the strong anthropic principle (that things are the way they are because otherwise we would not exist).

Marshall offered a solution to the paradox of "adaptive evolution", discovered by the British physician John Cairns ("The origin of mutants", 1988): some bacteria can mutate very quickly, way too quickly for Darwin's theory to be true. If all genes mutated at that pace, they would mostly produce mutations that cannot survive. What drives evolution is natural selection, which prunes each generation of mutations. But natural selection does not have the time to operate on the very rapid mutations of these bacteria. There must be another force at work that "selects" only the mutations that are useful for survival. Marshall thinks that the other force is the wave function's tendency towards choosing states of life and consciousness. Each mutation is inherently biased towards success.

His wife, the US philosopher Danah Zohar, expanded on his theory. Zohar views the theory of Bose-Einstein condensation as a means to reduce mind/body dualism to wave/particle dualism: the wave aspect of nature yields the "mental" (conscious experience), whereas the particle aspect of nature yields the material.

Zohar is fascinated by the behavior of bosons. Particles divide into fermions (such as electrons, protons, neutrons) and bosons (photons, gravitons, gluons). Bosons are particles of "relationship", because they are used by other particles to interact. When two systems interact (electricity, gravitation or whatever), they exchange bosons. Fermions are well-defined individual entities, just like large-scale matter is. But bosons can completely merge and become one entity, more like conscious states do. Zohar claims that bosons are the basis for the conscious life, and fermions for the material life.

The properties of matter arise from the properties of fermions. Matter is solid because fermions cannot merge. Likewise, she thinks that the properties of the conscious mind arise from the properties of bosons, because bosons can share the same state and they are about relationships.

This would also explain how there can be a "self". The brain changes all the time and therefore the "self" is never the same. I am never myself again. How can there be a "self"? Zohar thinks that the self does change all the time, but quantum interference makes each new self sprout from the old selves. Wave functions of past selves overlap with the wave function of the current self. Through this "quantum memory" each new "quantum self" reincarnates past selves.

The Ubiquity of Consciousness

The US physicist Nick Herbert thinks that consciousness is a pervasive process in nature. Consciousness is as fundamental a component of the universe as elementary particles and forces. The conscious mind can be detected by three features of quantum theory: randomness, "thinglessness" (objects acquire attributes only once they are observed) and interconnectedness (John Bell's discovery that, once two particles have interacted, they remain connected). Herbert thinks that these three features of inert matter can account for three basic features of our conscious mind:

free will, essential ambiguity, and deep psychic "connectedness". Scientists may be vastly underestimating the quantity of consciousness in the universe.

The US computer scientist James Culbertson speculated that consciousness may be a relativistic feature of space-time. He, too, thinks that consciousness permeates all of nature, so that every object has a degree of consciousness.

According to Relativity, our lives are world lines in space-time. Space-time does not happen: it always exists. It is our brain that shows us a movie of matter evolving in time.

All space-time events are conscious: they are conscious of other space-time events. The "experience" of a space-time event is static, a frozen region of space-time events. All the subjective features of the "psycho-space" of an observer derive from the objective features of the region of space-time that the observer is connected to. Special circuits in our brain create the impression of a time flow, of a time-travel through the region of space-time events connected to the brain.

Memory of an event is re-experiencing that space-time event, which is fixed in space-time. We don't store an event, we only keep a link to it. Conscious memory is not in the brain: it is in space-time.

The inner life of a system is its space-time history. To clarify his view, Culbertson presents the case of two robots. First a robot is built and learns German, then another robot is built which is identical to the first one. Culbertson claims that the second robot does not speak German, even if it is identical to the one that speaks German. Their space-time histories are different.

At the same time, Culbertson thinks that our consciousness is much more than an illusory travel through space-time, and it can, in turn, influence reality. Quantum Theory prescribes that reality be a sequence of random quantum jumps. Culbertson believes that they are not random but depend on the system's space-time history, i.e. on its inner life.

The Implicate Order
In the 1950s the US physicist David Bohm extended his ideas about the "implicate order" to the conscious mind.

Quantum and Relativity theories may be very different, but they agree in denying the existence of single static particles, they agree in describing the world as an undivided whole in constant flux (albeit in completely different ways) in which all parts of the universe are constantly interacting; and those parts include the observer, the "i". The universe is characterized by a "flow" that integrates everything: individual forms are the equivalent of the still frames of an object in motion.

It turns out that we perceive the "flow" of reality through those static images, but those still images are only a simplification of motion. By analogy, what goes on in our mind is a stream of consciousness, from which we can abstract concepts, ideas, etc (forms of thought) that are mere instances of that flow of thought. Thought is a kind of movement, and concepts are kinds of objects. Bohm believed that there is just one flow, in which both matter and mind flow, and that this flow can be known only implicitly through the forms (the still frames) that we can grasp out of this flow.

Bohm rejected the distinction between what we are thinking and what is going on, as well as the notion that one part of reality (my conscious mind) can know another part of reality: it is wrong to separate the thinker from the thought. The thinker is not separate from the reality that he thinks about, the thinker and that reality are parts of the same flow.

The belief that thinker and the object of his thinking (between thought and non-thought) are separate permeates our conscious life. This conviction is even built into the structure of language itself: modern language is based on the pattern "subject- verb- object", that clearly separates the subject and the object. But in reality the key actor is the verb, not the subject, and the verb unites the subject and the object in one, undivided action.

At the level of the "implicate order", which is a sort of "higher dimension", there is no difference between matter and mind. That difference arises within the "explicate order" (the conventional space-time of Physics). As we travel inwards, we travel towards that higher dimension, the implicate order, in which mind and matter are the same. As we travel outwards, we travel towards the explicate order in which subject and object are separate.

There is an inherent affinity between consciousness and implicate order. For example, when we listen to music we directly perceive the implicate order, not just the explicate order of those sounds.

Bohm's quantum field contains "active in-formation" that determines what happens to the particle ("in-formation" as in "give form"). Bohm interpreted the "active in-formation" of the quantum field that, in his view, accompanies each particle, as the "mental" (proto-conscious) property of the particle. Every particle has a rudimentary "mind-like" quality. Matter has "mental" properties, as well as physical properties. In-formation turns out to be the bridge between the two worlds. The two sides cannot be separated because they are entangled in the same quantum field. At the lower level of reality, mental (conscious) and physical processes are essentially the same.

The Mental State Of Particles

David Bohm's quantum field contains "active information": the form of the field determines the energy and momentum of a particle. Active information is information that is relevant to determining the movement of a particle. This information is not an opinion: it is the objective aspect of reality. Bohm interpreted this kind of information as the "mental" property of particles. Thus matter has "mental" properties as well as physical properties.

A quantum level mediates between an underlying mental level and the physical level that we observe. Bohm turns the table on supervenience: it is not the mental that supervenes upon the physical, it is the brain states that are dependent on mental states.

On top of Bohm's theory, the Finnish philosopher Paavo Pylkkanen offered a semiotic theory of the particle: the quantum field of the particle can be regarded as containing information about the environment surrounding the particle. The form of the wave function reflects the form of the environment. This form, in turn, determines the trajectory of the particle. The quantum field is a form of representation. Semiotically speaking (and using Charles Sanders Peirce's terminology), the quantum field is the sign, the environment is the object and the trajectory is the "interpretant".

Tripartite Idealism

The US physicist Henry Stapp elaborated on ideas already advanced in previous decades by John Von Neumann and Eugene Wigner that, basically, consciousness creates reality.

Stapp's theory of consciousness is grounded in Heisenberg's interpretation of Quantum Mechanics, that reality is a sequence of collapses of wave functions, i.e. of quantum discontinuities. Of all interpretations of Quantum Theory, this is also the closest to William James's view of the mental life as "experienced sense objects".

Stapp's view harks back to the heydays of Quantum Theory, when it was clear to its founders that "science is what we know". Science specifies rules that connect bits of knowledge. Each of us is a "knower" and our joint knowledge of the universe is the subject of Science. Quantum Theory is therefore a "knowledge-based" discipline. This view was "pragmatic" because it prescribed how to perform experiments, and it separated the system to be observed from the observer and from the instrument.

Von Neumann introduced an "ontological" approach to this knowledge-based discipline, which brought the observer and the instrument into the state of the system. Stapp describes Von Neumann's view of Quantum Theory through a simple definition: "the state of the universe is an objective compendium of subjective knowings". This statement describes the fact that the state of the universe is represented by a wave function which is a compendium of all the wave functions that each of us can cause to collapse with our observations. That is why it is a collection of subjective acts, although an objective one.

Stapp basically achieves a new form of idealism: all that exists is that subjective knowledge. Therefore the universe is not about matter: it is about subjective experience. Quantum Theory does not talk about matter, it talks about our act of perceiving matter. Stapp rediscovers George Berkeley's idealism: we only know our perceptions (observations).

Stapp's model of consciousness is tripartite. Reality is a sequence of discrete events in the brain. Each event translates into an increase of knowledge. That knowledge comes from observing "systems". Each event is driven by three processes that operate together:

- The "Schroedinger process" is a mechanical, deterministic, process that predicts the state of the system in a fashion similar to Newton's Physics: given its state at a given time, we can use equations to calculate its state at a different time. The only difference is that Schroedinger's equations describe the state of a system as a set of possibilities, whereas Newton's equations described it as just one certainty.

- The "Heisenberg process" is a conscious choice that we make when we decide to perform an observation. The formalism of Quantum Theory implies that we can know something only when we ask Nature a question. This implies, in turn, that we have a degree of control over Nature. Depending on which question we ask or not ask, we can affect the state of the universe. Stapp mentions the "Zeno effect" as a well known process in which we can alter the course of the universe by asking questions (it is the phenomenon by which a system is "frozen" if we keep observing the same observable very rapidly). We have to make a conscious decision about which question to ask Nature (which observable to observe). Otherwise nothing is going to "happen".

- The "Dirac process" gives the answer to our question. Nature replies with an "observed" quantity, and, as far as we can tell, the answer is totally random. Once Nature has replied, we have learned something: we have increased our knowledge. This is a change in the state of the universe, which directly corresponds to a change in the state of our brain. Technically, there occurs a reduction of the wave function compatible with the fact that has been learned.

Stapp's interpretation of Quantum Theory is that there are many knowers. Each knower's act of knowledge (each individual increment of knowledge) results in a new state of the universe. One person's increment of knowledge changes the state of the entire universe, and, of course, it changes it for everybody else too.

Quantum Theory is not about the behavior of matter, but about our knowledge of such behavior.

"Thinking" is a sequence of events of knowing, driven by those three processes.

Instead of dualism or materialism, one is faced with a sort of interactive "trialism", all aspects of which are actually mind-like. First, the physical aspect of Nature (the Schroedinger equation) is a compendium of subjective knowledge. Second, the conscious act of asking a question is what drives the actual transition from one state to another, i.e. the evolution of the universe. And, finally, there is a choice made from the outside, the reply of Nature, which, as far as we can tell, is random.

Stapp revives idealism by showing that Quantum Theory is about knowledge, not matter. The universe is a repository of knowledge, that we have access to and upon which our consciousness has control.

Quantum Idealism

Von Neumann's analysis of the collapse of the wave is also the basis for idealist theories that border on Eastern philosophy. For example, the Indian physicist Amit Goswami criticizes materialism (the view that consciousness is a material phenomenon, and that matter is the only substance) and endorses idealism (the view that matter is a mental phenomenon, and consciousness is the only substance).

Goswami's idealism is based on a simple postulate: that consciousness collapses the quantum wave, as Von Neumann originally claimed. Goswami claims that this brand of idealism has no problem with some notorious quantum "paradoxes". Basically, the "oddities" of Quantum Mechanics are in our mind, not in the world. For example, Schroedinger's cat could not possibly be both alive and dead in the real world, but it can be in our minds.

Paul Wigner extended Von Neumann's meditation by asking: if a friend tells me what she observed, at which point did the wave collapse, when the friend carried out her observation or when I carried out the observation of my friend telling me the result of her observation? If one thinks that the friend collapsed the wave, then the problem is that two friends observing the same phenomenon would both collapse the wave, and possibly observe opposite outcomes. If one thinks that I collapse the wave when I listen to my friend, then my friend's knowledge depends on her talking to me.

Goswami explains Paul Wigner's dilemma by arguing that there is only one consciousness, only one subject, and not many individual, separate subjects. There is, ultimately, only one observer. He also cites the non-locality of Quantum Mechanics as evidence that there is only one consciousness in the universe. Goswami credits consciousness with a deliberate act of determining reality: consciousness "chooses" (not just picks up) the outcome of a measurement.

Goswami thinks that a mind is made of ideas or thoughts, and he envisions these mental objects as fully equivalent to the material objects (the particles) studied by Quantum Mechanics. Thus they must obey the same Physics, i.e. the same theories about uncertainty, measurement and non-locality. He views the brain as both a quantum system and a measuring apparatus. The possibilities of the quantum brain are unconscious processes, and the equivalent of the physical observation is the one unconscious process that becomes conscious. Consciousness "chooses" which unconscious process becomes conscious (just like it "chooses" the outcome of any other measurement). Consciousness "chooses" its own conscious experience.

The Illusion of the Body

The US physicist Fred-Alan Wolf argued that the source of matter is conscious mind. The conscious mind "invents" a fictitious body and then it starts believing that "it is" the body.

Wolf makes reality arise from the limitations that Quantum Theory imposes on the human mind: we cannot ever know the exact position of a particle, therefore the particle is a purely mental hypothesis, therefore it exists only because our mind cannot ever know all about it. If we extend this line of reasoning to all matter, we reach the conclusion that the entire world that we perceive is an illusion, and that illusion is due to the fact that our mind cannot know the world as it really is. Reality has to do with perception of reality. If nobody observes it, it does not exist.

If reality is created by the observer, where is the observer? Wolf claims that the observer is not in the brain. His conclusion is that the observer, by observing, becomes the body: the observed and the observer are the same thing. After all, it is the observer who creates the physical world: that also includes the observer's own body.

The many-verse model of Quantum Theory states that all possible alternatives of a quantum system actually take place, one in each possible world, and that the observer splits in as many observers as possible worlds. Each observer in each world observes only one of the many possible realities. The world and the observer in it keep splitting as more possibilities arise. So conscious mind and matter get intertwined into story lines. Each story line is a memory of a past. Everything is alive because everything has a story line, which is both mind and matter. The story lines of a complex object form a "braid". The braid of story lines in a human body is a "script". A script is the collection of all the stories told by all the cells of the body. Bodies are scripts. Each cell is both matter and mind. The conscious mind is all over the body.

Wolf does not believe that Darwinian evolution alone can account for the birth of life and the working of natural selection. He believes that additional information is needed to start the mechanism of life, and that information must be coming from the future. Based on the same ideas from Quantum Theory (that reality exists only insofar as somebody observes it), Wolf claims that Nature produced the right organisms to survive in their environment because information flowed back from the future to the present about which organisms made sense. An observer can change the past by fixing the outcome of an observation: this would determine the past events that led to that outcome. Every time we "fix" the outcome of an observation, we force a certain past on the object of our observation. Our conscious mind can create a past from all the possible pasts. (This is yet another variation on the "Zeno effect": the life of a particle depends on how many times we observe it, because each observation changes its state).

Space, time, matter and consciousness are tied together by Relativity and Quantum Theory: Relativity binds together space-time and matter, while Quantum Theory binds together matter and consciousness.

Zero-point Consciousness

Another popular candidate for a quantum theory of consciousness, formulated notably by the Hungarian philosopher Erwin Laszlo, is the zero-point energy, the energy of the vacuum, which, according to Quantum Physics, is not zero because fluctuations always occur. More than 99% of all physical "matter" consists of a vacuum, and this yields a huge amount of zero-point energy. Laszlo believes that David Bohm's "in-formation" (an active form of information that shapes the informing agent) as the fundamental constituent of the universe expressed through a ubiquitous field that originates from the quantum vacuum. All universes originate in the quantum vacuum and evolve thanks to it. The vacuum generates a holographic field that encodes everything in spacetime. The Big Bang did not create it, but it amplified its fluctuations. Consciousness too is created by the quantum vacuum, and it too permeates the entire universe. Everything in the universe is interconnected thanks to that "Akashic" field.

The holographic field not only connects things but creates coherence. The intensity of the connection (and mutual influence) between two things is proportional to how similar the two things are. We are "in-formed" more by other humans than by, say, trees; but, ultimately, we are in-formed by everything in the universe, and the holographic field created by the quantum vacuum is the mediator of such interactions. We are connected in a superconsciousness that evokes Bohm's "implicate order", Carl Jung's "collective unconscious" and Pierre Teilhard de Chardin's "noosphere". Since the holographic field is a memory of everything in spacetime, it is also a memory of our selves. This means that our experience is eternal. Our experience becomes part of the universe, and in-forms others.

In general, the US physicist William Tiller postulated the existence of a class of natural phenomena called "subtle energies", beyond the four fundamental forces, which are supposed to act on consciousness.

Holonomic Consciousness
Both David Bohm and the Austrian psychologist Karl Pribram advocated the hologram as a paradigm to explain the unity and holistic property of consciousness. A hologram is another product of a quantum phenomenon: it arises from information carried by a laser beam, which can be viewed as a particular kind of Bose-Einstein condensate.

The brain stores information in a distributed manner that provides for fault tolerance and for "cue-based" retrieval. It is fault-tolerant because damage to one portion of the information does not cause damage to the information as a whole; and it is cue-based because information can be retrieved based on just partial information.

Pribram believes that the brain organizes information by interference patterns just like a hologram. Holography, invented in 1948 by the British physicist Dennis Gabor, employs coherent beams of light. A hologram is a permanent record of the interference between two waves of coherent light. Each part of the hologram contains each part of the interfering waves. This means that each part of the hologram contains the entire image. The entire hologram contains more details about the image, but the image is present in every part of the hologram. When re-illuminated with one of the original coherent lights, a three-dimensional image appears.

It turns out that the storage capacity of holograms is enormous.

Pribram's "holonomic" model of memory relies on the fact that many properties of the brain are shared by holograms.

In Pribram's opinion a sensory perception is transformed into a "brain wave", i.e. into a pattern of electro-magnetic activation that propagates through the brain just like the wave-front in a liquid. This crossing of the brain provides the interpretation of the sensory perception in the form of a "memory wave", which in turn crosses the brain. The various waves that travel through the brain can interfere. The interference of a memory wave and a perceptual (e.g., visual) wave generates a structure that resembles a hologram.

Pribram believes that the same equations used by Gabor to develop holography are used by the brain to analyze sensory data. He showed that all perceptions (and not only colors or sounds) can be analyzed into their component frequencies of oscillation and therefore treated by Fourier analysis, a well known mathematical tool.

Quantum Brain Dynamics

The Heisenberg and Von Neumann tradition viewed the brain as a "quantum measuring device". But the Japanese physicist Kunjo Yasue claims that brain substrates uphold second-order quantum fields, which cannot be treated as mere measuring devices.

Yasue, building on the quantum field theory developed in the 1960s by the Japanese physicist Hiroomi Umezawa and on his concept of "corticons" as more primitive than "neurons", developed a "quantum neurophysics" to explain how the classical world can originate from quantum processes in the brain. He showed that brain dynamics can be represented by a "brain wave equation" similar to Schroedinger's wave equation.

Yasue thinks that several layers of the brain can host quantum processes, whose quantum properties explain consciousness and cognition. Yasue presents the brain as a macroscopic quantum system. He focuses on water mega-molecules in the space between neurons, which can combine to form extended quantum systems, interacting with the neural networks. He also focuses on the sensory system, whose quantum field causes some special molecules in the membrane of the neuron to undergo Froehlich condensation and cause, in turn, macroscopic coherence.

He focuses on structures such as microtubules which lie inside the neuron, and which contain quasi-crystalline water molecules that again lend themselves to quantum effects. The function of this quantum field could be cognitive: some particular quantum states could record memory.

Yasue focuses on a bioplasma of charged particles that interact with the electromagnetic field, an ideal vehicle for a merger of the sensory quantum field with the memory quantum field, an ideal vehicle for the creation of classical reality. He argues that classical order can continually unfold in this bioplasma.

Yasue shows how consciousness could arise from the interaction between the electromagnetic field and molecular fields of water and protein. Furthermore, Yasue maintains that the evolution of the neural wave function is not random, as would result from the traditional quantum theories, but optimized under a principle of "least neural action". Random effects of consciousness are replaced by a "cybernetic" consciousness that is more in the tradition of the self as a free-willing agent.

Yasue is not a connectionist. The fact that neurons are organized inside the brain is of negligible importance in his theory.

Quantum-gravitational Consciousness
Quantum effects at the level of the protein were studied by Michael Conrad ("Quantum Molecular Computing", 1992), who argued that the molecules inside each cell might be implementing a kind of quantum associative memory. The protein is, after all, a biomolecular information processing system ("Information processing in molecular systems", 1972).

The British physicist Roger Penrose believes that consciousness must be a quantum phenomenon and that neurons are too big to account for consciousness. The US biologist Stuart Hameroff provided a better candidate: the "cytoskeleton". Inside neurons there is a "cytoskeleton", the structure that holds cells together, whose "microtubules" (hollow protein cylinders 25-nanometers in diameter) control the function of synapses. Penrose believes that consciousness is a manifestation of the "quantum cytoskeletal state" and its interplay between quantum and classical levels of activity. (Penrose implicitly attributes a special status only to the microtubules that are in the brain, but they are also ubiquitous among cells in the rest of the body and the same quantum argument could apply for microtubules in the foot).

Reality emerges from the collapse or reduction of the wave function. But Penrose makes a distinction between "subjective" and "objective" reduction. Subjective reduction is what happens when an observer measures a quantity in a quantum system: the system is not in any specific state (the system is in a "superposition" of possible states) until it is observed, and the observation causes the system to reduce (or "collapse") to a specific state. This is the only reduction known to traditional Quantum Theory. Objective reduction is, instead, a Penrose discovery, part of his attempt to unify Relativity Theory and Quantum Theory.

Superposed states each have their own space-time geometry. Under special circumstances, which microtubules are suitable for, the separation of space-time geometry of the superposed states (i.e., the "warping" of these space-times) reaches a point (the quantum gravity threshold) where the system must choose one state. The system must then spontaneously and abruptly collapse to that one state. So, objective reduction is a type of collapse of the wave function that occurs when the universe must choose between significantly different space-time geometries.

This "self-collapse" results in particular "conformational states" that regulate neural processes. These conformational states can interact with neighboring states to represent, propagate and process information. Each self-collapse corresponds to a discrete conscious event. Sequences of events then give rise to a "stream" of consciousness. Proteins somehow "tune" the objective reduction which is thus self-organized, or "orchestrated".

In other words, the quantum phenomenon of objective reduction controls the operation of the brain through its effects on coherent flows inside microtubules of the cytoskeleton.

In general, the collapse of the wave function is what gives the laws of nature a non-algorithmic element. Otherwise we would simply be machines and we would have no consciousness.

Penrose and Hameroff believe that "protoconscious" information is encoded in space-time geometry at the fundamental Planck scale and that a self-organizing Planck-scale process results in consciousness

Basically, Penrose believes in a Platonic scenario of conscious states that exist in a world of their own, and to which our minds have access. However, Penrose's "world of ideas" is a physicist's world: quantum spin networks encode proto-conscious states and different configurations of quantum spin geometry represent varieties of conscious experience. Access to these states (consciousness as we know it) originates when a self-organizing process (the objective reduction), somehow coupled with neural activity, collapses quantum wave functions at Planck-scale geometry.

There is a separate mental world, but it is grounded in the physical world.

Consciousness is the bridge between the brain and space-time geometry.

The New Materialism: Naturalistic Dualism

The Australian philosopher David Chalmers believes that consciousness is due to "protoconscious" properties that must be ubiquitous in matter and that "psychophysical" laws, not of the "reductionist" kind that Physics employs, will account for how conscious experience arise from those properties. There is, instead, nothing mysterious about our cognitive faculties, such as learning and remembering: they can be explained by the physical sciences the same way these sciences explained physical phenomena.

In a sense, Chalmers changed the scope of the mind-body problem, by enlarging the "body" to include the brain and its cognitive processes, and by restricting "mind" to conscious experience. Cognition migrates to the body. Consciousness, on the other hand, is truly a different substance, or, better, a different set of properties, and cannot indeed be explained by the "natural" laws of the physical sciences. The study of consciousness requires a different set of laws, because consciousness is due to a different set of properties.

Chalmers contends that mental (or, better, brain) activity is more than just conscious experience. States of the brain cause behavior. For example, I drink because I am thirsty, I move my hands because I want to grab an object, I buy a plane ticket because I believe the fare will go up. These "mental" states may or may not be conscious. Chalmers therefore distinguishes between the conscious experience, that he calls the "phenomenal properties of the mind", and the mental states that cause behavior, that he calls the "psychological properties of the mind" (that is "cognition"). In other words, phenomenal states deal with the first-person aspect of the "mental", whereas psychological states deal with the third-person aspect of the mental.

Psychological properties have, by his definition, a "causal" role in determining behavior. Whether a psychological state is also a phenomenal (conscious) state does not matter from the point of view of behavior. What conscious states do is not clear, but we know that they exist because we "feel" them.

Mental properties can therefore be divided into psychological properties and phenomenal properties. These two sets can be studied separately. It turns out that psychological properties (such as learning and remembering) have been and are studied by a multitude of disciplines, such as Biology and Neurology, and in a fashion not too different from physical properties of matter (given their "causal" nature), whereas phenomenal properties constitute the "hard" problem. A psychological property causes some behavior, no less than most material properties. A phenomenal property is a fuzzier object altogether.

Chalmers also distinguishes awareness and consciousness: awareness is the "psychological" aspect of consciousness. Whenever we are aware, we also have access to information about the object we are aware of. Awareness is that access. It is a psychological state that has a "causal" nature. "Consciousness" is a term more appropriately reserved for the phenomenal aspect of consciousness (for the emotion, for the feeling).

Chalmers is, de facto, separating the study of cognition from the study of consciousness. Cognition is a psychological fact, consciousness is a phenomenal fact. Psychological facts, by virtue of their causal (or functional) nature, can actually be explained by the physical sciences. It is not clear, instead, what science is necessary to explain consciousness. To start with, Chalmers focuses on the notion of supervenience.

Chalmers goes to a great extent to clarify the theory of supervenience. A set Y of properties supervenes on a set X of properties if any two systems that are identical by properties X are also identical by properties Y. For example, biological properties supervene on physical properties: any two identical physical systems are also identical biological systems.

"Logical" supervenience (loosely, "possibility") is a variant of supervenience: some systems could exist in another world (are "logically" possible), but do not exist in our world (are "naturally" impossible). Elephants with wings are logically possible, but not naturally possible. Systems that are naturally possible are also logically possible, but not viceversa. For example, any situation that violates the laws of nature is logically possible but not naturally possible.

Natural supervenience occurs when two sets of properties are systematically and precisely correlated in the natural world. Logical supervenience implies natural supervenience, but not viceversa. In other words, there may be worlds in which two properties are not related the way they are in our world, and therefore two naturally supervenient systems may not be logically supervenient.

Chalmers then argues that most facts supervene logically on the physical facts: if they are identical physical systems, then they are identical, period. There are few exceptions and consciousness is one of them. Consciousness is not logically supervenient on the physical.

Thus Chalmers concludes that consciousness "cannot" be explained by the physical sciences (more appropriately, cannot be explained "reductively"). But Chalmers does not conclude that consciousness cannot be explained tout court: only that it cannot be explained the way the physical sciences explain everything else, i.e. by reducing the system to ever smaller parts. Chalmers leaves the door open for a "nonreductive" explanation of consciousness.

Chalmers does not rule out "monism", the theory that there is only one substance; he only rules out that the one substance of this world is matter as we know it with the properties we currently know.

Chalmers' theory of consciousness is a variant of "property dualism": there are no two substances (mental and physical), there is only one substance, but that substance has two separate sets of properties, one physical and one mental. Conscious experience is due to the mental properties. The physical sciences have studied only the physical properties. The physical sciences study macroscopic properties like "temperature" that are due to microscopic properties such as the physical properties of particles. Chalmers advocates a science that studies the "protophenomenal properties" of microscopic matter that can yield the macroscopic phenomenon of consciousness.

Electromagnetism could not be explained by "reducing" electromagnetic phenomena to the known properties of matter: it was explained when scientists introduced a whole new set of properties (and related laws): the properties of microscopic matter that yield the macroscopic phenomenon of electromagnetism.

Similarly, consciousness cannot be explained by the physical laws of the known properties but requires a new set of "psychophysical" laws that deal with "protophenomenal properties". Consciousness supervenes naturally on the physical: the "psychophysical" laws will explain this supervenience, i.e. they will explain how conscious experiences depend on physical processes.

Chalmers emphasizes that this applies only to consciousness. Cognition is governed by the known laws of the physical sciences.

Chalmers then turns to the relationship between cognition and consciousness. Phenomenal (conscious) experience is not an abstract phenomenon: it is directly related to our psychological experience. Consciousness interacts with cognition and that interactaction gets expressed via what Chalmers calls "phenomenal judgements" ("I am afraid", "I see", "I am suffering"). These are acts that belong to our psychological life (to cognition) but are about our phenomenal life (consciousness).

Chalmers is faced with a paradox: phenomenal judgements, that are about consciousness, belong to cognitive life, therefore can be explained reductively, but he just proved that consciousness cannot be explained reductively. The way out of the paradox is to assume that consciousness is not relevant, that we can explain phenomenal judgements even if/when we cannot explain the conscious experience they are about, i.e. the explanation does not depend on "that" conscious experience, i.e. "that" feeling or emotion is irrelevant.

Chalmers cautions that this conclusion does not necessarily imply that consciousness (as in "free will") is irrelevant for behavior, but it surely does smell that way. If we can explain behavior about consciousness without explaining consciousness, it is hard to believe that behavior requires consciousness.

Chalmers takes these facts literally: our statements about consciousness are part of our cognitive life, and therefore can be explained quite naturally, just like any other behavior. I speak about my feelings the same way i raise a hand. There is a physical process that explains why i do both. It also happens that we "are" conscious, not just that we talk about it, and that part cannot be explained (yet). If we had a detailed understanding of the brain, we could predict when someone would utter the words "I feel pain". So Chalmers believes that our talk about consciousness will be explained just like any other cognitive process, just like any other bodily process. This is not the same as explaining the conscious feelings themselves, and it leaves open the option that feelings are but an accessory, an evolutionary accident, a by-product of our cognitive life with no direct relevance to our actions.

Chalmers also offers an explanation of phenomenal judgement based on the theory of information. After all, his definition of "cognition" is pretty much that of "information processing": cognition is the processing of information, from the moment it is acquired by the senses to the moment it is turned into bodily movement.

Information is what "pattern" is from the inside. Consciousness is information about the pattern of the self. Information becomes therefore the link between the physical and the conscious.

Since information is ubiquitous, he also gets entangled in the question whether everything has feelings. If experience is ultimately due to information, there is no reason why anything would not be associated with "experience". Just like every other physical property that we know to be widespread in the universe, there is no reason why "experience" (defined as the macroscopic effect of "protophenomenal properties") should not be widespread. Objects that implement an information-processing system may well have a degree of consciousness. Chalmers' "natural dualism" is therefore a close relative of "panpsychism".

Furthermore, if information leads to experience, there must be a lot more experience than we "feel" because the brain processes a lot more information than we are aware of. But then parts of the brain may have experience that does not travel to the "i". The "i" is not necessarily all that is experienced by the brain. The "i" may simply be a chunk of coherent information out of the many that arise all the time in the brain.

Ultimately, David Chalmers believes that it makes no sense to talk of pieces of consciousness. Consciousness "is" the experience of being the subject, so, by definition, it is a unity: it is all of which I am the subject at a certain time. This has implications for any theory of consciousness, because the reductionist approach (splitting the problem into smaller problems) is, by definition, doomed to failure: consciousness cannot be split lest you lose precisely consciousness, and then you are no longer analyzing consciousness. Consciousness can only be studied as the state of being the subject, which is fundamentally different from the study of how the brain is enabled to integrate different processes. What is needed is a holistic approach to consciousness.

Panpsychism

The Roman philosopher Lucretius of the first century B.C. once observed that "every creature with senses is made only of particles without senses". The paradox still stands. Descartes did not solve it by simply separating "sense" (conscious mind) from "non-sense" (matter) and subsequent philosophers did not solve it no matter how they looked at the relationship between mind and matter.

The solution to the paradox has always been around, and it only required accepting that our mind is nothing else than a natural phenomenon.

For example, in the 1920s the British mathematician Alfred Whitehead argued that every elementary constituent of the universe must be an event having both an objective aspect of matter and a subjective aspect of experience. Some material compounds, such as the brain, create the unity of experience that we call "mind". Most material compounds are limited in their experience to the experience of their constituents.

The US philosopher Thomas Nagel ("Panpsychism", 1979) reached a similar conclusion: that "proto-mental properties" must be present in all matter, and, suitably organized, become somebody's consciousness. He believed in one, common source for both the material and the mental aspects of the world. Mental and material are never separated: there is never the material without the mental, and there is never the mental without the material.

The Danish physicist Niels Bohr once suggested that the quantum wave function of matter represents its mental aspect, that the wave of the electron is the equivalent of the mind of matter. Niels Bohr suggested that the duality of waves and particles could explain the duality of mind and matter. After all, the wave of probability could be interpreted as expressing a "free will" of the electron, a primitive "mental life" of its own. The dual aspect of body and mind within an organism would derive from the dual aspect of the particles composing an organism, from the dual aspect of wave and particle.

Panpsychism (the notion that everything is conscious to some extent) is the simplest way to explain why some beings (e.g., me) are conscious. After all we don't wonder why we are made of electrons: everything is made of electrons, therefore no wonder that my body too is made of electrons. We wonder why we are conscious because we made the assumption that only some things (us) are conscious. All we have to do is remove that assumption and we have a simple theory of consciousness.

Panpsychism has attracted followers from many disciplines: philosophers such as Hermann Lotze, in "Microcosmos" (1864), and Charles Hartshorne, in "Beyond Humanism" (1937); physicists such as Ernst Mach, in "The Analysis of Sensations and the Relation of the Physical to the Psychical" (1886), and David Bohm, in "A New Theory of the Relationship of Mind and Matter" (1986); mathematicians such as William Clifford, in "Body and Mind" (1874), and Alfred North Whitehead, in "Process and Reality" (1929); biologists such as Ernst Haeckel, in "Our Monism" (1892), and John Haldane, in "The Inequality of Man" (1932); theologians such as Pierre Teilhard de Chardin, in "Phenomenon of Man" (1959), and David Ray Griffin, in "Unsnarling the

World-Knot" (1998); psychiatrists such as William James, in "A Pluralistic Universe" (1909), and Theodore Ziehen, in "Epistemology of Psychophysiological and Physical Foundations" (1913), in which he named the fundamental constituents of consciousness "gignomena"; as well as a polymath such as Charles Peirce, in "Man's Glassy Essence" (1892).

The Austrian philosopher Karl Popper, instead, criticized panpsychism. First of all, Popper thinks that there is no need to panic: something does get created ("emerges") out of nothing all the time. Properties emerge at higher levels of organization that did not exist at the lower levels. Hence consciousness could just be an emerging property just like the properties of a liquid emerge from molecules that don't have does properties. Secondly, Popper thinks that there is a fallacy in the very notion that panpsychism would comply with the "nothing comes out of nothing" dogma: if the constituents of matter have a lower degree of consciousness than humans do, then their combination must somehow create higher and higher degrees of consciousness, and that "is" a case of something that comes out of nothing (that higher degree of consciousness is not present in the constituents).

A General Property of Matter

The Italian mathematician Piero Scaruffi offered his variant on panpsychism ("A simple theory of consciousness", 2001).

I am conscious. i am made of cells. Cells are made of molecules. Molecules are made of atoms, and atoms are made of elementary particles. If elementary particles are not conscious, how is it possible that many of them, assembled in molecules and cells and organs, eventually yield a conscious being like me?

Many attempts have been made at explaining consciousness by reducing it to something else. To no avail. There is no way that our sensations can be explained in terms of particles. So, how does consciousness arise in matter? Maybe it doesn't arise, it is always there.

No matter how detailed an account is provided of the neural processes that led to an action (say, a smile), that account will never explain where the feeling associated to that action (say, happiness) came from. No theory of the brain can explain why and how consciousness happens, if it assumes that consciousness is somehow created by some neural entity that is completely different in structure, function and behavior from our feelings.

From a logical standpoint, the only way out of this dead-end is to accept that consciousness must be a property of the particles that make up my body.

When we try to explain consciousness by reducing it to electrochemical processes, we put ourselves in a situation similar to a scientist who has decided to explain electrical phenomena by using gravity. Electrical phenomena can be explained only if we assume that electricity comes from a fundamental property of matter (i.e. from a property that is present in all matter starting from the most fundamental constituents) and that, under special circumstances, enables a particular configuration of matter to exhibit "electricity".

Similarly, if consciousness comes from a fundamental property of matter (from a property that is present in all matter starting from the most fundamental constituents), then, and only then, we can study why and how, under special circumstances, that property enables a particular configuration of matter (e.g., the human brain) to exhibit "consciousness".

Any paradigm that tries to manufacture consciousness out of something else is doomed to failure. Things don't just happen. Ex nihilo nihil fit. Consciousness cannot simply originate from the act of putting unconscious neurons together. It doesn't appear like magic. Conductivity seems to appear by magic in some configurations of matter (e.g. metallic objects), but there's no magic: just a fundamental property of matter, the electrical charge, which is present in every single particle of this universe, a property which is mostly useless but that under the proper circumstances yields the phenomenon known as conductivity.

Particles are not conductors by themselves, just like they are not conscious, and most things made of particles (wood, plastic, glass, etc. etc.) are not conductors (and maybe have no consciousness), but each single particle in the universe has an electrical charge and each single particle in the universe has a property, say, C. That property C is the one that allows our brain to be conscious. It is not that each single particle is conscious or that each single piece of matter in the universe is conscious. But each single particle has this property C which, under the special circumstances of our brain configuration (and maybe other brain configurations and maybe even things with no brain) yields consciousness.

Just like electricity and gravitation are macroscopic properties that are caused by microscopic properties of the constituents, so consciousness may be a macroscopic property of our brain that is caused by a microscopic property of its constituents. Just like electrical phenomena can only be reduced to smaller-scale electrical phenomena (all the way down to the electrical charge of each single constituent), so consciousness can only be reduced to smaller-scale conscious phenomena.

Property C has not been found by Physics for the simple reason that Physics was not built to find it: Physics is an offshoot of Descartes' dualism, which strictly separated mind and matter and assigned Physics to matter. Newton's Physics was built to explain the motion of bodies, and that is what it explains. It did not find elementary particles and it did not find entropy. It was built to explain bodies. Relativity was built to explain the constant speed of light, electromagnetism and gravitation. And that is what it explains. It did not find quarks either, because it was not built to study atoms. Quantum Theory, on the other hand, found quarks, because it was built to study the atom. But it did not find black holes, because it was not built to study gravitation.

Scaruffi's theory is neither dualistic nor materialistic. Like dualists, he admits the existence of consciousness as separate from the physical properties of matter as we know them; but at the same time, like materialists, he considers consciousness as arising from a "physical" property (that we have not discovered yet) that behaves in a fundamentally different way from the other physical properties. So in a sense it is not a "physical" property, but it is still a property of all matter. His is an identity theory, in that he thinks that mental corresponds to neural states, but it goes beyond identity because I also think that the property yielding consciousness is common to all matter, whether it performs neural activity or not.

What made Descartes believe in dualism is the unity of consciousness. But electrical conductors also exhibit a unity of electricity, and nonetheless electrical phenomena can be reduced to a physical property of matter.

The main problem is the lack of an empirical test for consciousness. We cannot know whether a being is conscious or not. We cannot "measure" its consciousness. We cannot rule out that every object in the universe, including each elementary particle, has consciousness: we just cannot detect it. Even when I accept that other human beings are conscious a) I base my assumption on similarity of behavior, not on an actual "observation" of their consciousness; and b) I somehow sense that some people (poets and philosophers, for example) may be more conscious than other people (lawyers and doctors, for example).

The trouble is that our mind is capable only of observing conscious phenomena at its own level and within itself. Our mind is capable of observing only one conscious phenomenon: itself.

A good way to start would be to analyze why consciousness is limited to the brain. Why does consciousness apply only to the brain? What is so special about the brain that cannot be found anywhere else? If the brain is made of ordinary matter, of well-known constituents, what is it that turns that matter conscious when it is configured as a brain, but not when it is configured as a foot?

A Reductionist Theory of the Self

The "i" is the central problem of consciousness. Even if we eventually explain how conscious experience arises from the electrochemical processes of the brain, even if we discovered some kind of "proto-consciousness" that gets combined to form emotions and feelings, we will still need to show how and why that set of emotions and feelings becomes an "i".

That matter can feel emotions is mystery enough. But what we feel is even stranger: it is not that each part of our body, each molecule feels emotions. It is "i" who feels those emotions.

A body is made of parts that interact, and each one has its own life. But a consciousness is an "i" that feels all of the emotions related to that body. My consciousness is not distributed the same way that matter is distributed in my body.

Let us assume that everything is conscious to some degree. Every atom, every molecule, every tissue, every organ, every being is "conscious". And that "i" is just what i am conscious of. If i were born a finger, i would only be conscious of what a finger does. "I" happen to be born the part of the brain that is conscious of what i am conscious.

In this scenario of multiple "consciousnesses", the "i" that is writing this sentence is not the conscious part of the foot or the nail, it is the conscious part of a part of the brain. I am not conscious of my foot's consciousness, because "i" am the consciousness of something else (a part of the brain).

And i am not conscious of the consciousness of any other parts of the brain because "i" (the one who is writing right now) am not those parts. "I" am the consciousness of a part of the brain, and it turns out that "i" (this particular consciousness) receive information from several parts of the body and direct order to several parts of the body. For example, it is likely that the "consciousnesses" associated with my fingers are conscious of typing on the keyboard. They have to, because the brain tells them to. "I" (the consciousness of that part of the brain) am only aware of sending them the order to type. Because of the organization of the body, the brain controls other organs. Because each part is associated with a consciousness, each part is aware of what it is doing. But "i", the consciousness associated with this part of the brain, identify with the whole body.

"I" am actually not conscious of everything. There is a consciousness associated with my liver and one with my intestine and one with each of millions of minuscule parts. "I" am not aware of any of those parts.

I cannot feel that i because "i" am not that i

Evolution has decreed which parts are connected to the part of the brain that is associated with "i". "I" am aware only of those parts. The main difference between "i" and other "consciousnesses" inside this body is that "i" (or the brain part associated with that "i") can tell fingers to type this sentence, and can tell the mouth to utter words.

Of course, there could be another consciousness (another i) that is conscious of parts of my body that i am not conscious of. Maybe there is a consciousness (another i) that is directing me to think what i am thinking and directing me to direct the fingers to type what they are typing. I cannot be conscious of this "super-consciousness" or of any other consciousness associated with my body, because "i" am not it or them.

The reason things do not get out of control is that the structure of "consciousnesses" must mirror the structure of the body, so that an order issued by a consciousness cannot conflict with the order issued by another consciousness.

Finally, in this scenario of multiple consciousnesses, it is not necessary that "consciousnesses" be truly directing anything. Each consciousness could simply be the "phenomenal" aspect of a physical process: any action by a body part also yields a conscious experience that presumes being the cause of that action. Whether it is consciousness that directs the body or viceversa is another issue.

The body is made of parts, each part being made of parts and so forth all the way down to elementary particles. Each of those parts, all the way down to elementary particles, also has a phenomenal aspect, i.e. its own "consciousness".

The consciousness that is writing this sentence is "i". There are countless consciousnesses that share this body, each of them conscious of what one part of the body is doing. They may all be convinced of having free will, just like I am. There may be consciousnesses that share this body and direct "i" to do what "i" am doing.

In this scenario, "i" cannot feel any other consciousness than "i", because that is what "i" am.

Further Reading
Bohm, David: WHOLENESS AND THE IMPLICATE ORDER (Routledge, 1980)
Bohr, Niels: ATOMIC THEORY AND THE DESCRIPTION OF NATURE (1934)
Chalmers, David: THE CONSCIOUS MIND (Oxford University Press, 1996)
Chalmers, David: THE CHARACTER OF CONSCIOUSNESS (Oxford University Press, 2010)
Culbertson, James: THE MINDS OF ROBOTS (University of Illinois Press, 1963)

Culbertson, James: SENSATIONS MEMORIES AND THE FLOW OF TIME (Cromwell Press, 1976)
De Quincey, Christian: "Radical Nature" (Invisible Cities, 2002)
Eccles, John: EVOLUTION OF THE BRAIN (Routledge, 1989)
Gardenfors, Peter: HOW HOMO BECAME SAPIENS (Oxford Univ Press, 2003)
Goswami, Amit: THE SELF-AWARE UNIVERSE (Putnam, 1993)
Griffin, David: "Unsnarling the World-Knot" (University of California Press, 1998)
Herbert, Nick: ELEMENTAL MIND (Dutton, 1993)
Laszlo, Erwin: SCIENCE AND THE AKASHIC FIELD (Inner Traditions, 2004)
Lockwood, Michael: MIND, BRAIN AND THE QUANTUM (Basil Blackwell, 1989)
Marshall, I.N., Zohar, Danah: QUANTUM SOCIETY (William Morrow, 1994)
Mathews, Freya: "For Love of Matter - A Contemporary Panpsycism" (SUNY, 2003)
Nagel, Thomas: MORTAL QUESTIONS (Cambridge Univ Press, 1979)
Penrose, Roger: THE EMPEROR'S NEW MIND (Oxford Univ Press, 1989)
Penrose, Roger: SHADOWS OF THE MIND (Oxford University Press, 1994)
Penrose, Roger and Hameroff, Stuart: CONSCIOUSNESS AND THE UNIVERSE (Journal of Cosmology vol 14, 2011)
Popper, Karl & Eccles, John: THE SELF AND ITS BRAIN (Springer, 1977)
Pribram, Karl: LANGUAGES OF THE BRAIN (Prentice Hall, 1971)
Pribram, Karl: BRAIN AND PERCEPTION (Lawrence Erlbaum, 1990)
Pylkkanen, Paavo: MIND, MATTER AND ACTIVE INFORMATION (Univ. of Helsinki, 1992)
Rosenblum, Bruce and Kuttner, Fred: QUANTUM ENIGMA (Oxford Univ Press, 2006)
Searle, John: THE REDISCOVERY OF THE MIND (MIT Press, 1992)
Skrbina, David: PANPSYCHISM IN THE WEST (2005)
Stapp, Henry: MIND, MATTER AND QUANTUM MECHANICS (Springer-Verlag, 1993)
Tiller, William: SCIENCE AND HUMAN TRANSFORMATION (1997)

Von Neumann, John: DIE MATHEMATISCHE GRUNDLAGEN DER QUANTENMECHANIK/ MATHEMATICAL FOUNDATIONS OF QUANTUM MECHANICS (Princeton University Press, 1932)
Walker, Evan: THE PHYSICS OF CONSCIOUSNESS (Perseus, 2000)
Whitehead, Alfred: PROCESS AND REALITY (1929)
Wolf, Fred Alan: MIND INTO MATTER (Moment Point, 2001)
Yasue, Kunio & Jibu, Mari: QUANTUM BRAIN DYNAMICS AND CONSCIOUSNESS (John Benjamins, 1995)
Zohar, Danah: QUANTUM SELF (William Morrow, 1990)

THE SELF AND FREE WILL: DO WE THINK OR ARE WE THOUGHT?

The Self

Consciousness is more than just being aware of being. It comes with a strong notion: the distinction between self and non-self. I know that i am myself, but i also know that i am not anybody else, and that nobody else is me. I know that i am myself, and i know that i was myself yesterday and the day before and the year before and forty years ago. Consciousness carries a sense of identity, of me being me. And it comes with a sense that there are other selves.

Differentiation of self and the other is a fundamental property of living organisms. Even plants use protein discrimination mechanisms, and most organisms could not survive without the ability to distinguish alien organisms.

The US neurologist Roger Sperry can be said to have founded the scientific study of the self, when ("Mind, Brain, and Humanist Values", 1965) he posited that the self must be an "emergent" property of brain processes that, in turn, controls brain processes. A similar theory ("downward causation") was advanced by the US psychologist Donald Campbell ("Downward causation in hierarchically organised biological systems", 1974).

This emergent property, consciousness, is thus generated by brain (neural) processes but, once it is born, it is no longer a brain (neural) phenomenon: it belongs to a different category that does not obey neurological laws anymore ("non-reductive physicalism"). Sperry believed that there is only one substance (as in monism). However, entities of that substance can create new entities that exhibit completely different properties, just like Quantum Mechanics tells us that interactions among elementary particles can result in (emergent) phenomena that have properties apparently unrelated to the properties of the particles. In general, Sperry thought that this is the way that human values emerge from the physical structure of our body, and they too constitute a completely independent category of entities.

The physical stuff of our body (and of our brain) changes all the time. Cells die and are replaced, old synapses die and new synapses are created. The US mathematician Norbert Wiener concluded that the self cannot possibly be a material entity, that identity cannot possibly be due to physical continuity, and argued that, instead, the self must be a pattern of organization, "patterns that perpetuate themselves".

A preliminary question is where the self comes from. We inherit bodily traits from our parents: do we also inherit the consciousness of our self from our parents' selves? After all, my brain's structure is probably very related to my mother's and/or my father's brain. If the self (my feeling of who i am) is due to the processes inside my brain, then my self should be somehow similar to the self of my parents.

The Narrative Self

The US psychologist Jerome Bruner believed that narratives are important for the creation of the self.

We have a strong feeling that we are a particular "i" (our identity): where does it come from? At the same time, we are capable of turning sensory input into a "narrative": we not only catalog all the images, sounds, etc that we perceive, we also organize them in "stories". It appears that there is a biological need to "make sense" of our experience, and to structure that sense into "narratives". Narratives seem to link our current status to past events and future actions.

One particular case of narrative is the "autobiography": the story about myself. Is that the cause or the effect of the "self"?

Another particular case of narratives is constituted by the narratives about others: as we organize their actions in stories, we construct theories of their minds, of why they do what they do. This separates the self from the non-self, and places the self in relationship with other selves.

Narratives are, inevitably, subjective. They do not, and do not intend to, "duplicate" reality: they internalize reality, they interpret reality from the vantage point of the self. In a sense, therefore, our narratives "falsify" experience. In fact, the self is a "perpetually rewritten story". The self that we remember is the one we need to survive today. If that self does not "work" anymore, we introduce a turning point in the narrative that changes our self.

Bruner believes in a multiplicity of narratives. The only way that one can fuse the different chronological selves of a life (from childhood to present) is by telling a story: all those selves become characters of the story, the same way a novelist uses several characters to create a plot. At the same time, the story that one fabricates is heavily influenced by the stories that one has heard. One's culture creates the templates that one uses in creating new stories. There is no single, static remembered self. What we remember is influenced by social and cultural factors. Self-narratives do not even depend so much on memory as on thinking.

Bruner thinks that narratives are not only an accident of nature but play an important role in creating our understanding of the community and of ourselves. In other words, Bruner believes that "making sense" (i.e., constructing meaning) is the fundamental characteristic of our self-conscious life.

The Illusion of the Self

However, the self is not a simple concept. The US biologist Ulric Neisser identified five kinds of self-knowledge: the ecological self (situated in the environment), the "interpersonal self" (situated in the society of selves), both based on perception, the private self, the conceptual self and the narrative (or "remembered") self.

Worse: we might overestimate the self and its presumed free will. The US neurophysiologist Benjamin Libet discovered that the will to act follows the act ("Production of threshold levels of conscious sensation by electrical stimulation of human somatosensory cortex", 1964), and in later experiments he estimated that the "readiness potential" precedes movement by about half a second, and awareness of this "decision to act" follows by about 300 milliseconds. The motor cortex activates 500 milliseconds before we are aware that we want to carry out an action. In other words, the brain decides unconsciously to act, before we are aware of having decided to act. We become aware of the action only if the neural event lasts about 500 milliseconds. A way to interpret this is that we are conscious only of electric field patterns in the brain that last about half a second. The scary notion, of course, is that a) we are not conscious of many "decisions" that our brain makes (anything that occurs inside our brain in less than half a second) and b) we are conscious of "our" decisions only "after" the brain has already decided them. This can be interpreted as proof that free will is an illusion, or that free will has about 200 milliseconds to "veto" what the brain wants to do. Libet's experiments showed, however, that, to wit, the brain is ahead of the mind.

By the same token, the US psychologist Daniel Wegner showed that it is relatively easy to trick people into believing that they decided to do something when in fact someone else had; a fact that to him proved the "illusion of control".

Robert Ornstein thinks that different regions of the brain behave independently of consciousness, as proved by the fact that sometimes consciousness realizes what has been decided after it has already happened. The self is only a part of our "mental" life, and not always connected with the rest of it. The self shares the brain with other selves. Selves take hold of consciousness depending on the needs of the moment. Each self tends to run the organism for as long as possible. The self only occasionally notices what is going on. Continuity of the self is an illusion: we are not the same person all the time. Different selves within the brain fight for control over the next action.

Michael Gazzaniga believes that the self is only an "interpreter" of the conclusions that the various brain processes have reached.

Daniel Dennett also has difficulties with the self. In his "multiple draft" theory, the self is simply the feeling of the overall brain activity. Whichever draft, whichever "narrative" dominates is my current "i". But the dominant draft could be changing every second. Dennett is opposed to the idea that there is an enduring self because it would imply that there is a place in the brain where that self resides. He thinks that such "Cartesian theater" is absurd and that our conscious life is implemented by multiple parallel drafts.

The British philosopher Derek Parfit believes that the self "is" the brain state. As the brain state changes all the time, the self cannot be the same: there cannot be a permanent "self". "i" do not exist. What exists is a brain state that right now is "me". The next brain state will also be a self, distinct from the previous one. There is a chain of successive selves, each somehow linked through memory to the previous one. Each self is distinct from the previous ones and future ones. The "i" is a mere illusion. There is no person that all those selves share. Derek Parfit believes in a Buddhist-like set of potential "consciousnesses", each with its own flow of feelings, although at each time the one which dominates gives me the illusion of having only one consciousness and one identity.

The US neurophysiologist Paul Nunez interprets Libet's experiments as showing that a) it takes about half a second to become aware of something because that awareness is due to some global brain activity with a lot of loops; and b) the conscious and the unconscious are in continuous communication, a feedback loop of its own.

The Solipsistic Brain

The US neurophysiologist Walter Freeman discovered that the neural activity due to sensory stimuli disappears in the cortex and in lieu of it an apparently unrelated pattern appears, as if the brain created its own version of what happens in the world. Most of the sensory input is basically wasted. Freeman came to believe that "a form of epistemological solipsism isolates brains from the world". The brain creates patterns that have little or nothing to do with the real world: these patterns yield a world that is consistent and complete, based on (basically) computational efficiency, not on accuracy. Each brain creates its own world, which is internally consistent and complete.

Contrary to a popular paradigm, perception does not consist of information reception, processing, storage, and recall. Perception is the creation of meaning, a very "subjective" process.

These "solipsistic" brains communicate, basically, by "unlearning": unlearning is a process by which a brain must give up its beliefs and learn new ones through "socially cooperative" actions.

The Autobiographical Self

The US psychologist John Kotre focused on autobiographical memory: memory of the infinite sequence of details that creates the story of the self. When we think of ourselves in the distant past, we are often part of the memory: we can see ourselves in the scene. Thus at some point the remembering self (the self as subject) fashions a remembered self (the self as object).

A clue comes from memories that are more vivid than the average. These vivid memories tend to be of three kinds: novel, consequential and emotional events. But there is also a fourth kind of vivid memory, the one that Ulrich Neisser calls "repisode": the symbolic episode that summarizes several preceding episodes. Sometimes a new event stands as a symbol for some pattern that has been going on for years. For example, a gentle gesture by an old friend is a reminder that this friend has always been there to help when needed. The repisode "makes sense" of many previous episodes.

Like Gazzaniga, Kotre believes that the self is due to an "interpreter": it remembers itself as the center of things and makes sense of everything else. This interpreter is both a librarian, who simply archives memories, and a "myth-maker", who creates the myth of the "i".

Kotre points out that the youngest children cannot attribute to others their own qualities. For example, a four-year old boy cannot grasp the idea that his brother has a brother. But one or two years later this becomes obvious. Thus at a certain stage in cognitive development we develop the concept of the "me": we as we are seen by someone else. This opening up of the perspective enables the child to tell stories that are not just self-centered. The "i" is beginning to shape the "me". By the end of the teens, the child has acquired the ability to reflect on herself. The child (now no longer a child) has acquired the ability to create a myth of herself. In the rest of her life, the adult simply continues to refine that "myth". The main task of this myth-making process is to create the sense of continuity: however different my body was when I was a child, that was the same "I" that is now writing these lines. As we develop our myth of ourselves, we also change the memories of ourselves: a memory from the distant past is inevitably affected by the myth we have created of ourselves. The "myth-maker" is as relevant as the "keeper of archives" for the purpose of reconstructing memories. As we get older, the keeper of archives recedes because literal truth is no longer needed, and the myth-maker becomes more and more important in shaping our memories. Thus memories become more mythic and less accurate, but the sense of the self is still preserved. In fact it may be better preserved this way.

The Grounded Self

On the other hand, the existence of a unitary and continuous self is defended by the US psychologist Richard Carlson. He believes that the self is a biological feature.

Following James Gibson and Ulrich Neisser, Carlson thinks that every act of perception specifies both a perceiving self and a perceived object. Seeing something is not only seeing that object: it is also seeing it from a certain perspective. The perspective implies a "seer". There is no act of perception of an object without a subject. The subject is as much part of perception as the object.

This co-specification of self and object is useful for adding the "first person" to the information-processing paradigm of mental processes, which cannot traditionally deal with the self.

Carlson's fundamental move is to distinguish between the content and the object of a mental state: content and object are not the same thing, because the content includes both the object and the subject. The "mode" of content specifies the self, the "aboutness" of content specifies the object (the environment).

Following John Searle's analysis of speech acts, Carlson further distinguishes between the content of a mental state and the "noncontent" of that mental state. The "noncontent" includes the purpose of the mental state (for example, the degree of commitment), and even its "implementation" properties (for example, the duration of the state, etc). A mental state has content and non-content, and non-content is as important as content.

This analysis serves to elucidate that there is more than just an object in an act of perception. There is more than just a scene in a visual perception: there is a subject that is seeing, there is a purpose of seeing (for example, "i am spying" versus "i am gazing") and there is a duration.

Contrary to Dennett and Gazzaniga, Carlson reaches the conclusion that the continuity of consciousness is not only real, but it is even an ecological necessity, because the self is co-specified by perception, and perception is driven by changes in the world, and those changes are continuous. Cognition is grounded in one's point of view, and that point of view is grounded in an environment, and this two-fold grounding process is continuous.

A World Apart

The Austrian philosopher Karl Popper thinks that there can be no consciousness without memory, without a process that provides some continuity of memory, linking one conscious act with other conscious acts.

Popper argues that you don't learn about your self by self-observation but by the process of constructing your self. Initially, as a child, you do so by interacting with others in a preverbal way; and later through language. However, Popper emphasizes that you are not just a passive receiver of knowledge, but an active searcher and builder of knowledge, including about yourself. You construct a theory of who you are. The self, in fact, consists of all the theories that you create about the universe. You actively explore the world and build theories about its various aspects. Popper rejects the idea that the self preexists experience and is the thing that experiences: it is experience that creates the self, not viceversa.

The self is, first and foremost, a sense of being an individual distinct from other individuals. Popper views this as an extension of the fact that life tends to create individuals, not clones. Popper speculates that this fact is the very cause of the emergence of mind and consciousness. The biological fact that all individuals are different explains why a need emerged for a conscious mind. The unity of the self is a consequence of "biological individuation". Popper thinks that, without individuation, living beings would not have had a need for a mind to drive their behavior.

Bodies change over the course of a lifetime. So do minds, that learn and forget. Metabolism of the body proceeds in parallel with "metabolism" of the self, but the sense of identity is retained through all the physical and mental changes. However, there is an asymmetry between the two: the body comes first, the self develops later. You are first a body, eating, screaming, gesturing; and only later do you become a conscious self. Popper believes in an intermediate stage, a stage in which the child discovers that she is a person, not a thing; and then this "person" evolves into a full-fledged conscious "self". Once created, the self is permanently active, exploring the world, and creating theories about the world. Those theories get "selected" by the experience of the world, refuted or refined, in an endless process of trial and error. Once created, the self drives the acquisition of new knowledge, which can happen consciously or unconsciously, but always by interaction with the environment and by the related autonomous process of theory formation and refinement. All our non-innate knowledge comes from such a process: a theory of how things work endows us with a set of expectations; experience selects whether an expectation has to be retained or erased; experience produces new theories, with new sets of expectations; experience modifies those theories so that the expectations due to them will match reality.

Popper thinks that three worlds coexist: the physical world (World 1), the mental world (World 2) and the world of products of mental activities (World 3). The main biological function of World 2 (mind, self) is to produce theories and expectations; and the main biological function of World 3 is to make these theories vulnerable to the judgment of experience. World 3 is basically the place where we simulate the outcome of planned behavior without risking our lives. World 3 emerged during evolution to provide a safer way to evolve.

The way in which the objective knowledge of World 3 is created is very similar to how natural selection creates species. Natural selection acts on biological traits, a similar selection process acts on behaviors (which are initially programmed by genes), and a similar selection process acts on knowledge (which is initially transmitted by culture). At all three levels (World 1, 2 and 3, i.e. physical bodies, behaviors and theories) two forces fight each other: instruction is the conservative force (that creates and tries to preserve biological traits, behaviors and theories) while selection is the revolutionary force (that introduces new traits, behaviors and theories).

At the same time Popper notes that our learning process is driven by our expectations, because the expectations drive our behavior, and therefore shape our exploration of the world, and therefore determine our experiences, which in turn refine our theories and our expectations.

One particular theory that we develop over the course of our life is the theory of who we are: the self itself is a theory that gets refined via a process of trial and error, via a process of "natural selection". The self is the very active process of creating theories and expectations, and of integrating all our theories and expectations, The developing plan of our life "is" the self that gives unity to our mental life. Popper's metaphor is that the self is the active programmer and the brain is the passive computer.

Identity: Who Am I?

Every year 98% of the atoms of my body are replaced: how can I claim to be still the same person that I was last year, or, worse, ten years ago? About 70% of your original neurons die before you reach maturity: what is the relationship between you and the child that had all those dead neurons? What is (where lies) my identity? What is "my" relationship to the metabolism of my body?

Derek Parfit once proposed this thought problem: what happens to a person who is destroyed by a scanner in London and rebuilt cell by cell in New York by a replicator that has received infinitely detailed information from the scanner about the state of each single cell, including all of the person's memories? Is the person still the same person? Or did the person die in London? What makes a person such a person: bodily or psychological continuity? If a person's matter is replaced cell by cell with equivalent cells is the person still the same person? If a person's psychological state (memory, beliefs, emotions and everything) is replaced with an equivalent psychological state is the person still the same person? What if the original is not destroyed, and now there is a perfectly identical copy of yourself living in London? Are both you?

Parfit believes that teleportation is simply a way for the self to travel from New York to London: yes, you are the same person at the other end. On the other hand, you die all the time because your brain changes all the time, and your self now is not the same self of a few minutes ago. And since you died all the time that you thought you were living, the final death is not a big deal: it is just one more death in a long sequence of deaths, or, better, it is yet another self that dies.

This is not just philosophy for the sake of philosophy: your morality depends on what you think "you" is. If you think of yourself as a continuous being, you reach some conclusions. If you think of yourself as identical to your brain, and therefore dying and being born another person all the time, you reach different conclusions on what is right and what is wrong. Your rational and emotional behavior depend on that assumption.

The most obvious paradox is: how can reality be still the same as we grow up? Do two completely different brains see the same image when they are presented with the same object? If the brains are different, then the pattern of neural excitement created by seeing that object will be different in the two brains. How can two different brains yield the same image? The logical conclusion is "no, the tree I see is not the tree you see, we just happen to refer to it the same way so it is not important what exactly we see when we look at it". But then how can I see the same image yesterday, today and tomorrow? Our brain changes all the time. Between my brain of when i was five years old and my brain of today there is probably nothing in common: every single cell has changed, connections have changed, the physical shape of the brain has changed. The same object causes a different neural pattern in my brain today than it did in my brain forty years ago. Those are two different brains, made of different cells, organized in different ways: the two patterns are physically different.

Nonetheless, it appears to me that my old toys still look the same. But they shouldn't, because my brain changed, and the pattern they generate in my brain has changed: what i see today should be a different image than the one i saw as a five-year old. How is it that i see the same thing even if i have a wildly different brain?

Furthermore, experience molds the brain: i am not only my genome, i am also the world around me. And i change all the time according to what is happening in the world. "I am" what the world is doing.

All of this almost seems to prove that "i" am not in my brain, that there is something external to the brain that does not change over time, that the brain simply performs computations of the image but the ultimate "feeling" of that image is due to a "soul" that is external to the brain and does not depend on cells or connections.

On the other hand, it is easy to realize that what we see is not really what we think we see.

When we recognize something as something, we rarely see/feel/hear/touch again exactly the same thing we already saw/felt/heard/touched before. I recognize somebody's face, but that face cannot possibly be exactly the same image i saw last time: beard may have grown, a pimple may have appeared, hair may have been trimmed, a tan may have darkened the skin, or, quite simply, that face may be at a different angle (looking up, looking down, turned half way). I recognize a song, but the truth is that the same song never "sounds" the same: louder, softer, different speakers, static, different echo in the room, different position of my ear with respect to the speakers. I recognize that today the temperature is "cold", but if we measured the temperature to the tenth decimal digit it is unlikely that we would get the exact same number that i got the previous time i felt the same cold. What we "recognize" is obviously not a physical quantity: an image, a sound, a temperature never repeat themselves. What is it then that we recognize when we recognize a face, a song or a temperature? Broadly speaking, it is a concept.

We build concepts of our sensory experience, we store those concepts for future use, and we match the stored concepts with any new concept. When we do this comparison, we try to find similarity and identity. If the two concepts are similar enough, we assume that they are identical, that they are the same thing. If they are not similar enough, but they are more similar than the average, then we can probably establish that they belong to a common super-concept (they are both faces, but not the same face; they are both songs, but not the same song; and so forth). We have a vast array of concepts which are organized in a hierarchy with many levels of generalization ("your face" to "face of you and siblings" to "faces of that kind" to "generic face" to ... to "body part" to ...). A sensory experience is somehow translated into a concept and that concept is matched with existing concepts and eventually located at some level of the hierarchy of concepts. If it is close enough to an existing concept of that hierarchy at that level, it is recognized as the same concept. Whatever the specific mechanism, it is likely that what we recognize is not a physical quantity (distribution of colors, sound wave or temperature) but a concept, that somehow we build and compare with previously manufactured concepts.

Identity is probably a concept. I have built over the years a concept of myself. My physical substance changes all the time, but, as long as it still matches my concept of myself, I still recognize it as myself.

Are We Immortal?

Ultimately, it depends on one's definition of identity. If I build an exact copy of an object, is it the "same" object? In the case of inanimate matter, the temptation is to answer that a copy is just that: a "copy". But things assume a more sinister look when dealing with brains. If I build an exact copy of your brain, which presumably yields the same mental life as yours, is that copy "you"? Does identity require those specific atoms in that specific place, or only the same kind of atoms and the same way they are related?

It is likely that only a finite number of brains are possible, because brains are self-organizing systems, i.e. they are systems that tend to happen at certain "attractive" configurations while shunning many others. No matter how complex the brain is, there are probably only so many configurations that realize a stable system that works like a brain. In other words, some brains just cannot exist.

If that is true, then as long as life repeats itself on an Earth-like planet, "you" are likely to be eventually "rebuilt" again, i.e. to live again.

Assuming that the universe's life is infinite, i.e. that it will exist forever and ever, then the odds that it will again reach (somewhere sometimes) the conditions present on today's Earth are significant, and thus your brain, given enough time (which is in vast supply in eternity), is likely to be created again. Not only once, but infinite times.

Will that be really "you", or just something with the exact same genes and brain that you have now?

What Is A Self?

What makes you "you"? What would make you somebody else? If we transplant a brain from one body to another body, who is who? Is the person her brain or her body? Most people (and even most philosophers) would be reluctant to accept a brain transplant because they think that it would be an other brain which gets your body, not "you" who get a new brain. In other words, whatever "i" means, that thing is inside my brain, and it goes where my brain goes. If my brain is transplanted into the body of a supermodel, "i" have become a supermodel and her body is now "my" body. If the supermodel's brain is transplanted into my body, my body is now "her" body. The brain determines where "you" are.

The British philosopher Eric Olson is one who disagrees. Olson believes that identity comes from biology, not from psychology. You can be brain dead, but still be "you". For as long as some biological functions continue, Olson thinks that you are you. If someone transplants your brain to another body, you still are the same "you", and someone has received "your" brain. This, of course, flies in the face of the definition of "i". Olson thinks that psychological continuity is one thing, and identity is another thing. Psychological continuity occurs between two "people" at two different times: one at a certain instant is psychologically continuous with the other one at a later instant. In other words, Olson thinks that minds change all the time, and therefore i am no longer the mind that i was a second ago. Therefore it is improper to claim that "i" was. In a sense, "i" can only "be" now. "I am" is correct, whereas "i was" is a contradiction (the "i" that was is gone).

The US physician Lewis Thomas argued that even the most primitive organisms must have a sense of self. It is not about being smart: it is about "being". But then he also preached that the self is a myth, because we are part of a bigger self that eventually includes all life. The Earth is a giant self.

Conceptual Life

The US mathematician Douglas Hofstadter noted that the self can exist not only because of an ability (consciousness, which is basically a tautology) but also because of an inability: we are unable to perceive the working of our brain. Our mind's ability to create symbols (categories, concepts, ideas) out of the signals it receives through the senses becomes the very limit of our mind's ability to understand reality: it cannot perceive anything at lower levels. Our mind cannot peer below the level of symbols. Were we able to perceive the detailed operations of our brain's neurons, we probably wouldn't be "self-aware". The self is an illusion that is created by the fact that we do not perceive ourselves as billions of neurons exchanging electrochemical messages. We perceive ourselves and our lives and the world around us as concepts: sun, busy, good, tired, etc. At the physical level each of these is an electrochemical process in our brain that involves millions of neurons. We know it as a scientific fact, but we do not perceive it. Thinking is inherently opaque: when i think of myself, i think of my goals, failures, desires, memories, but i don't think of Neuron 345-769-045 triggering Neuron 745-809-760 triggering Neuron...

On the other hand, the self is capable of perceiving the symbolic activity: we "know" that we know something, we can retrace the steps of our logic, we can argue with ourselves why we believe in something. We cannot perceive the working of the neurons that make concepts possible (the working of our neural life), but we perceive the working of our conceptual life. That perception "is" us.

"Our very nature is such as to prevent us from fully understanding its very nature".

The Importance of Being Warm

When speculating about consciousness, identity and free will, it is important not to forget what bodies are and how they work.

Among the many features of living organisms, one is often overlooked: each living organism can live only within a very narrow range of temperature. Temperature is one of the most crucial survival factors.

Temperature also happens to be an important source of "identity". For example, water and ice are made of the same atoms: it's the temperature that determines whether you are water or you are ice.

It's the temperature that determines whether your body is dead or alive, and it's the temperature that determines whether you are lying and shivering in bed or are playing soccer outside. Our identity does change with the temperature of our body (from no identity to "regular" identity to delirious identity).

Most of what our body does has nothing to do with writing poems or making scientific discoveries: it is about maintaining a stable temperature.

Retrospection vs Introspection

There is a school of thought that denies the meaning of any such discussion about the self.

The British philosopher Gilbert Ryle argued that Descartes invented a myth when he provided definitions for the mental and the physical, as if they were two different things; when he assumed that every human is both a body (that is in space and is subject to the laws of Physics) and a mind (that is not in space and is not subject to the laws of Physics); that a person lives two parallel lives, one as a body and one as a mind (one being a public history and the other being a private history because nobody can witness your inner thoughts). Descartes created the myth of what Ryle parodies as "the ghost in the machine" (the ghost being the mind, the machine being the body). Because things in space tend to obey the law of cause and effect, then we tend to think that mind (which is not in space) too obeys the law of cause and effect. Because the physical world is deterministic, we tend to think that the mental world must be deterministic too. This leads to the belief that there is a machine called mind inside the machine called body, although the mind machine is significantly different from the body machine. Ryle believes that this view is based on a "categorical mistake", and that both Idealism and Materialism (in trying to reduce one realm to the other) fall in this categorical mistake.

First of all, Ryle takes aim at the notion that first we need to think before we can act. He points out that "efficient practice precedes the theory of it". A skilled craftsman does not need to think about how to do things, but he does need to think if asked. In fact, our behavior is mostly driven by habits, especially "expert" behavior (the most "intelligent" of all behaviors). Theorizing is not necessary to carry out intelligent actions. In some cases it is not even sufficient: an encyclopedic medical knowledge is not enough to be a good surgeon. "Knowing how" does not depend on "knowing that".

Furthermore, if that belief were true, there would be an infinite regression of "theorizing" because any theory of action is itself an operation whose execution must depend (according to this view) on thoughts about it. Intelligent behavior cannot possibly consist in first thinking of it and then executing it.

The interactions between mind and body are no less problematic. If such interactions are neither mental nor physical, they obey neither the laws of Psychology nor the laws of Physics. If that were true, Ryle argues that it would be impossible for people to understand other people. Understanding, instead, can be easily explained as a form of "knowing how". We understand a person's actions because we know how those actions work. Carrying out an action and understanding another person's actions are two sides of the same coin: it is the competence (the knowing how to perform that action) that enables one to both carry out the action and to understand another person performing that action. The rules that are at work in understanding another person's action are the same that one needs to perform that action. One does not infer the workings of another person's mind, one follows them.

The "will" is yet another complication. Ryle points out that, according to the "ghost in the machine" model, a mind lives in three modes: the cognitive mode, the emotional mode and the "conative" mode. Volition, however, has the same "infinite regression" problem of the cognitive mode: if a volition is "voluntary", than it must mean that it must have been thought, but then this thought must have been willed to, and so forth.

Ryle blames the concept of consciousness on the Protestant ethics, that required each person to keep track of her moral state without the aid of the Catholic confessor. This implied the existence of the capacity for introspection. At the same time Galileo's science was introducing the experimental method (and therefore the figure of the observer) to study the behavior of matter, so it came natural to introduce "consciousness" as the observer of the mental world.

Ryle attacks the widely-shared hypothesis that there exists both consciousness and self-consciousness (introspection), i.e. that a mind can observe its own mental working, i.e. that we are conscious of what is happening to us and at the same time able to introspect what we are conscious of. In reality, i cannot attend to two things at the same time. In fact, if i focus on what i am thinking, i am no longer thinking of "that". And very often focusing on the actions that are being performed changes the actions that i will perform. I cannot be laughing hysterically and observing myself laughing hysterically because i stop laughing hysterically when i observe myself. It is only in retrospect that i can observe myself laughing hysterically. I cannot be aware of daydreaming at the same time that i am daydreaming, but only a few seconds later, when i actually stopped daydreaming. Ryle does not deny that we somehow keep track of what we think. He only points out that thinking of what we are thinking is a paradox: either we are thinking of it or we are thinking of what we are thinking. We don't think of what we are thinking, we think of what we just thought: our introspection can follow our train of thoughts, but it will always be one step behind (just like you cannot jump on your shadow's head, and a missile cannot be its own target). There is no need for a monitoring process to know what we are thinking; and such a process would be impossible anyway. In reality, introspection is always "retrospection".

Ryle even attacks the view that I can access to my mental life but not to yours. Ryle argues that this is a myth too. Whatever we can know about ourselves we learn it the same way we learn what we know about others, namely by observing behavior. Self-knowledge and knowledge of others are therefore similar processes that only differ in degrees: sometimes we can know more about ourselves than about others, and sometimes we can know more about others than ourselves (because sometimes it is easier to face the truth about others than it is about ourselves). Both the knowing of others and the knowing of ourselves are notoriously imprecise: we often fail to appreciate our character the same way that sometimes others deceive us.

This act of cognition (of discovering one's mental life) consists not in a separate cognitive process but in a process very similar to the one that allows us to appreciate the skills of someone else because we ourselves have those skills. Otherwise we would fall into another trap of infinite recursion: if there is a cognitive process analyzing my cognitive process, then one could also envision a cognitive process analyzing the cognitive process of analyzing my cognitive process, and so forth ("an infinite number of onion-skins of consciousness embedding my mental state"). Instead, we asses our own and others' mental life by inferring predispositions and abilities to do something (just like the British constitution signifies the predisposition and ability of Great Britain to vote democratically, and from observing the British go to the polls one can infer that the British constitution prescribes that). We do not observe someone's mind but only a number of predispositions to act in a manner rather than in another. We can observe that someone is more or less selfish, more or less hot-tempered, etc. And the same applies to what we know about ourselves. We don't find it inside our mind but by observing our performance. The way to know if you know something is to test you, to listen to you telling me about it, and the way to know if i know something is to test myself (usually in silence). We are naturally equipped with the skill of assessing what lies behind one's actions, and even of discovering hypocrites and charlatans. This is exactly the same kind of process by which i understand your (or my own) physical abilities: by observing performance. Ultimately, the observation of what we consider to be mental life (our own or others') is an automatic process of theory formation by induction.

Ryle's view can be summarized as: mental processes cannot be isolated from physical processes, and, in fact, states of mind should better be viewed as actions of the body. Our vocabulary for someone's mental states is in reality a vocabulary for someone's predispositions or abilities or inclinations to perform some actions.

Purpose

Why do living things do what they do?

The purposefulness of living organisms is simply a consequence of evolution driven by variation and natural selection. Living organisms have a fundamental goal, survival, and have inherited a repertory of behaviors to achieve that goal. Survival ultimately depends on self-regulation.

The 19th-century French psychologist Claude Bernard "discovered" the self-regulating nature of living organisms. Bernard realized that each living organism is a system built to maintain a constant internal state in the face of changing external conditions. The regulation of this "milieu interieur" is life itself, because it is this stable state that gives the organism its independence from the environment, i.e. its identity. This is the dividing line that separates animate and inanimate matter: inanimate matter obeys Newton's laws of cause and effect, animate matter tends to maintain its state no matter what external forces are applied. Unlike objects, whose state is changed when a force is applied, the state of a living organism is not changed by an external force. The living organism, as long as it is alive, maintains its state constant.

The "purposeful" behavior of a living organism is the reaction to environmental forces: the organism needs to act in order to continuously restore its state. A body seems to "want", "intend", "desire" to maintain its internal state (either by eating, moving, sleeping, etc), a state that, ultimately, is a combination of chemical content and temperature. Living bodies appear to act purposefully, but they are simply reacting to the environment.

For Bernard "freedom" is independence from the environment. Control of the internal state allows a living organism to live in many different environments. The living organism is "free" in that is not a slave of its environment.

Bernard's idea of self-regulation extended to all living organisms. Humans are not the only ones to have "goals". Animate behavior "is" control of perception.

Life is Unpredictable
The fundamental problem of free will is how the determinism of brain matter turns into the freedom of the self.

Assuming that humans do have free will, do animals also have the same free will that humans have? Or are they only machines that move according to formulas?

There is no evidence that at any point in time one can predict the next move of a chicken or an ant. No matter how simple and unconscious animals seem to be, their behavior is still largely unpredictable. You can guess what the chicken will want to do, but you can never be sure, and you can never guess the exact movements. There are infinite paths an ant can follow to go back to the nest and the one it will follow cannot be predicted. At every point of that path the ant can choose where to go next. Two ants will follow two different paths. Each ant seems to have its own personality.

Even the movement of mono-cellular organisms is unpredictable to some extent. No matter how small and simple the organism, a degree of free will seems to be there. Free will seems to be a property of life. What triggers the next move of bacteria, ants and chicken is not just a Newtonian formula. If they are machines, then these machines do not obey classical Physics. There is a degree of freedom that every living organism seems to enjoy. And it doesn't require a sophisticated brain. There is a degree of freedom that just shouldn't be there, if Newton was right.

If these are machines, they are machines that cannot be explained with our Mechanics because at every point in time there are many possible time evolutions and all seem to be possible, and none can be exactly predicted.

Will, Not Necessarily Free: A Materialistic View Of Free Will

The problem with free will is that it does not fit too well with the scientific theories of the universe that have been developing over the last three centuries.

While those theories are fairly accurate in predicting all the natural phenomena we deal with, they don't leave much room for free will. Particles behave the way they behave because of the fundamental laws of nature and because of what the other particles are doing; not because they can decide what to do. Since we are, ultimately, collections of particles, our free will is an embarrassment to Physics.

On the other hand, a simple look at the behavior of even a fly seems to prove that free will is indeed a fact and is pervasive. Free will is a fundamental attribute of life. A robot that moved but only repeating a mechanical sequence of steps would not be considered "alive". Life has very much to do with unpredictability of behavior, not just with behavior. Or, better, behavior is behavior inasmuch as it is unpredictable to a degree; otherwise it is simply "motion".

Whether it is indeed "free" or not, "will" (the apparent ability of an ant to decide in which direction to move) appears to be an inherent feature of life, no matter how primitive life is. A theory of life that does not predict free will is not a good theory of life. Somehow, "free" will must be a product of the chemistry of life, at some very elementary level. In other words, obtaining the right chemical mix in the laboratory would not be enough: that mix must also exhibit the symptoms of free will.

The origin of "free" will, therefore, appears to be life itself.

The universe had a small entropy at the Big Bang moment. Free will is possible only because the past has a low entropy and the future has a high entropy. Life works against this process (it turns high entropy into low entropy) and therefore must be contributing to the increase in entropy in the environment (the overall entropy can never decrease). In a sense, life seems to be the very process that creates, sustains and increases free will.

Free Will

Is consciousness merely an "observer" of what is going on in the brain (of neural processes), or is consciousness a "creator" as well of neural processes?

Some scientists (Albert Einstein among them) argued that consciousness must be fabricated by reality, that what we feel is simply an unavoidable consequence of the state of the universe, that we are simply machines programmed by the rest of the universe.

Other scientists believe the opposite, that consciousness fabricates reality, that we have the power to alter the course of events. They believe in free will.

Do we think or are we thought?

This question is misleading. The question is, in a sense, already an answer: the moment we separate the "i" and the body, we have subscribed to dualism, to Descartes' view that spirit and matter are separate and spirit can control matter.

A free will grounded in matter is not easy to picture because we tend to believe in an "i" external to our body that controls our body.

But, in a materialist scenario, the "i" is supposed to be only the expression of brain processes. If that is the case, then "free will" is not about the "i" making a decision: the "i" will simply reflect that decision. What makes the decision is the brain process.

This does not mean that free will cannot exist. It just needs to be redefined: can a brain process occur that is not completely caused by other physical processes?

In a materialist scenario, free will does not require consciousness: consciousness is an aspect of the brain process that "thinks". The question is whether that brain process has free will.

If consciousness is indeed due to a physical process, if consciousness is ultimately material, does this preclude free will? For centuries we have considered free will to be an exclusive property of the soul, mainly because 1. We deemed the soul to be made of spirit and not matter, and 2. Nothing in Physics allows for free will of matter.

If we now recognize that consciousness is a property of matter (possibly one that occurs only in some special form and configuration of matter, but nonetheless ultimately matter), the second statement must be re-examined because the possibility of free will depends on its truth. If the motion of matter is controlled only by deterministic laws, then free will is an illusion. On the other hand, if matter has a degree of control over its own motion, then free will is a fact.

The question is not whether we have free will, but whether the laws of our universe (i.e., Physics) allow for free will.

Free Will and Randomness

Free will is often associated with randomness: a being has free will if it can perform "random" actions, as opposed to actions rigidly determined by the universal clockwork. In other words, free will can exist only if the laws of nature allow for some random solutions, solutions that can be arbitrarily chosen by our consciousness. If no randomness exists in nature, then every action (including our very conscious thoughts) is predetermined by a formula and free will cannot exist.

In their quest for the source of randomness in human free will, both neurologists like John Eccles and physicists like Roger Penrose have proposed that quantum effects are responsible for creating randomness in the processes of the human brain. Whether chance and free will can be equated (free will is supposed to lead to rational and deterministic decisions, not random ones) and whether Quantum Theory is the only possible source of randomness is debatable.

Free Will and Quantum States

We "feel" that we have free will (that we are in control of our decisions) but everything we find in science tells us the opposite: every movement of everything (including living beings and including sentient beings) is being explained by scientific formulas.

First of all, one has to define "free will". It is, in fact, easier to define the opposite of free will: an object does not have free will if all its movements are caused by laws of Physics (or, better, by "natural laws", since those laws were invented by Nature and not by physicists).

Then the real question is why would there be natural laws to start with. Why do all electrons obey the same electromagnetic laws? Why do all masses of water (all other things being the same) boil at the same temperature? Why do all members of a category behave the same way under some forces?

Could there be a universe in which the same force on two identical objects in identical conditions has completely different effects? Could there be a universe that is totally unpredictable? Could there be a univese in which the consequence of every action is determined by a throw of dice?

That's precisely how Quantum Physics describes our universe: whether a particle will be found in one place or another is completely random. There are constraints, but within those constraints the particle is "free" to be in any place.

One would be tempted to write: "The particle is free to be wherever it wants to be".

Free will of this kind is inherent in matter. All matter, at the fundamental level, has "free will".

The word "will", however, is usually employed to mean "conscious will". Does an electron consciously choose where to be when someone observes it? To me, that's a different question. We assume that "free will" has to be conscious, but why can't there be unconscious things with free will and conscious beings with no free will? Many people spend their entire life studying, working and even passing their time based on what society "markets" to them. Why can't there be a being that is perfectly conscious of what is happening but incapable of making any decision, basically an intelligent appliance?

Does our refrigerator have free will? No. Is it conscious of being a refrigerator? We cannot know. I don't think that the two questions should be asked at the same time. The definition of "conscious" and the definition of "free will" are not identical.

Assuming that our gut feeling is correct, we do have free will and our refrigerator probably does not have it. If the elementary particles that constitute all matter do have it, why is it that humans have it (to an even larger degree) and refrigerators lose it?

In most cases when several particles are combined in a system, their universe of possibilities (i.e. their "free will") gets reduced to just one value. The particle is in one specific place. It has lost all free will. It is as if the "free wills" of all the particles neutralize each other. The resulting system does not exhibit free will.

There might be cases in which combining particles amplifies (instead of neutralizing) their free will in a sort of positive feedback loop: then the free will of the whole system should be much greater than the free will of each particle. The free will of the whole system is actually a quantum effect, even if it is no longer recognized as a quantum effect because its behavior (like love, work, sport, investment) has nothing to do with the quantities that are studied at the quantum level (like position, momentum, energy, spin).

The human brain might just be such a system, a system that does the exact opposite of what it does to the things it observes. When it observes (i.e. interacts with) a particle, it "collapses" its wave of possibilities. When it observes itself (i.e. the particles of the brain interact with each other), it "explodes" the waves of possibilities into what we normally call "free will".

The Nonlinear Origins of Free Will

Creativity is a property of life. No insect nor worm moves in a predictable manner, and no insect or worm follows the same trajectory again if moved to the exact same starting position.

Memory is reconstructive: you never remember anything as it was. If you tell the same story over and over again, you will use different words all the time (it comes as an unnatural effort to memorize one particular sequence of words). Gestures and sentences are always improvised. If you perform the same action a thousand times, in exactly the same position under the exact same conditions, you will always perform a different sequence of movements.

The reason (or at least one reason) that we cannot repeat ourselves is not that the initial conditions change but that "we" change all the time: we are never the same again.

It is impossible to put "you" in the exact same initial conditions because "you" are the one element that is different even if everything else in the universe remains the same. Ditto for the ant, ditto for the worm. Anything that is alive changes all the time, therefore will never repeat itself.

If i knew the equations that govern your brain, i might indeed be able to calculate the trajectory that you will follow to go from here to there; but your brain will change the moment you start moving (in fact, every time you breathe and every time you absorb sunlight), which means that the equations change as you go.

You are creative and unpredictable because your brain is governed by a nonlinear equation.

Whether this can be called "free will" or not depends on definitions: it is unpredictable what "you" will do next.

The reason why the movement of insects seems to be so erratic (and therefore driven by free will) is that a tiny change makes a big difference on their nervous system, and, just like us, these organisms change with every particle of oxygen they breathe, with every photon that hits their eyes, with every food they digest.

Robots do not exhibit "free will" because their actions can be predicted, and they simply repeat the same action if the conditions are the same. The reason they behave in a repetitive manner is that they don't change while they exist. They are designed to remain the same, except for updating their knowledge of the state of the world. Their "nervous system" (their "self") does not change with every electrical impulse that they receive and with every photon that hits their sensors. They do not change most of the cells of their body during a year: they only change the components that fail, and even those get replaced with identical copies. Given the same conditions (the same state of the world) a robot's arm will indeed follow the exact same trajectory to grab an object and a robot's "mouth" will utter the exact same words to tell a story.

The problem of free will is framed incorrectly. The "i" that is supposed to have free will does not exist: it is something that changes all the time, because at every instant countless cells of the body change including countless cells of the brain.

Hence the "i" that is supposed to have free will is actually defined by that "free will": it is the sequence of unpredictable actions generated by a nonlinear system.

You yourself cannot predict what your free will will make you do and think in a few seconds, let alone a few years from now.

Free will exists, but the "i" does not exist.

The Machine's Free Will

Why do we claim that a machine has no free will? Usually, because a machine can solve only the problems that we program it to solve. We, on the other hand, can solve novel problems in unpredictable situations (or, at least, give them a try). And that's because we can make actions that we have never done before and that nobody ever told us to do, whereas a machine can only do what it has been programmed to do.

Machines are built to solve specific problems in specific situations, simply because that is what humans are good at: building machines that solve specific problems in specific situations: we humans like to "design" a machine, to write the "specifications", etc. This is not the way nature built us. Nature built us on a different principle and it is no surprise that we behave differently from machines. Since in nature we never know what the next problem and situation will be like, nature built us as "Darwinian" machines: our brains generate all the time a lot of possible actions and then pursue the ones that are "selected" by the environment (the specific problem and situation at hand). Nature built us on a different principle than the one we use to build machines. The main difference between our mind and a machine is their architectures.

The lack of free will in machines is not a limitation of machines: it is a limitation of our mind. If we built a machine the same way nature builds its cognitive beings, i.e. with the same type of architecture, it would be a rather different machine, capable of generating a huge amount of random behaviors and then picking the one that best matches the current problem and situation. One can even envision a day when machines built with a "Darwinian" architecture (descendants of today's genetic algorithms and neural networks) will "out-free will" us, will exhibit even more free will than we do. After all, most of the time we simply obey orders (we obey publicity when we shop, we obey record labels when we sing a tune, we obey our mother's moral principles all day long), whereas a machine would have no conditioning. And it may be able to generate a lot more alternatives than our brain does. Free will is simply a folk name for the Darwinian architecture of our mind that was created by Nature. When we choose an action, we are actually obeying an ancient command of Nature. We are simply machines of a different kind than the ones we build.

Do we think or are we thought?

Further Reading

Bruner Jerome: SELF RECONSIDERED (1995)
Carlson, Richard: EXPERIENCED COGNITION (Lawrence Erlbaum, 1997)
Dennett, Daniel: KINDS OF MINDS (Basic, 1998)
Freeman, Walter: MASS ACTION IN THE NERVOUS SYSTEM (Academic Press, 1975)
Hofstadter, Douglas: I AM A STRANGE LOOP (Basic, 2007)
Kotre, John: WHITE GLOVES (Norton, 1996)
Lazarus, Richard: EMOTION AND ADAPTATION (Oxford Univ Press, 1991)
Libet, Benjamin: MIND TIME (Harvard Univ Press, 2004)
Nunez, Paul: BRAIN, MIND, AND THE STRUCTURE OF REALITY (Oxford Univ Press, 2010)
Olson, Eric: THE HUMAN ANIMAL (Oxford Univ Press, 1997)
Ornstein, Robert: MULTIMIND (Houghton Mifflin, 1986)
Parfit, Derek: REASONS AND PERSONS (Oxford Univ Press, 1985)
Thomas, Lewis: la(Viking Press, 1974)
Ryle, Gilbert: THE CONCEPT OF MIND (Hutchinson, 1949)
Thomas, Lewis: THE LIVES OF A CELL (1974)
Wegner, Daniel: THE ILLUSION OF CONSCIOUS WILL (MIT Press, 2002)
Wiener, Norbert: HUMAN USE OF HUMAN BEINGS (1950)

FINALE: WHAT DOES IT ALL MEAN?

Epilogue: Piero Scaruffi's Theory Of Consciousness

I hope that this book is going to be useful, first and foremost, as a survey, so that many more people can be informed about research programs underway the world over. My own ideas, as presented mostly at the end of each chapter, can be summarized as follows.

First of all, two fundamental principles.

I believe in the existence of a common underlying principle that governs inanimate matter (the one studied by Physics), living matter (the one studied by Biology) and consciousness (studied by Cognitive Science).

We know that the world of living beings is a "Darwinian" system: competition, survival of the fittest, evolution and all that stuff. We know that the immune system is a Darwinian system. We are learning that the brain is also a Darwinian system, where the principles of natural selection apply to neural connections. It is intuitive that memory is a Darwinian system: we remember the things that we use frequently, we forget things we never use. I think that the mind is a Darwinian system as well: competition, survival of the fittest and evolution work among thoughts as well. The Darwinian system recurs at different levels of organization, and one of them happens to be our thought system, i.e. our mind.

Biology and Physics offer us different theories of Nature. Physics' view is "reductionist": the universe is made of galaxies, which are made of stars, which are made of particles. By studying the forces that operate on particles, one can understand the universe. Biology's view is Darwinist: systems evolve. Reconciling the two views is the great scientific challenge of the next century.

The main addition to the Darwinian paradigm that i advocate is a crucial role for endosymbiosis: I believe that new organisms can be created by "merging" two existing organisms. If each organism is made of smaller organisms, then it is not surprising that a Darwinian law governs each level of organization: each component organism "was" a living organism, and, like all living things, was designed to live and die and evolve according to the rules of Darwinian evolution. The organism that eventually arose through the progressive accretion of simpler organisms is a complex interplay of Darwinian systems. It is not surprising that muscles, memory, the immune system and the brain itself all exhibit Darwinian behavior (get stronger when used, weaker when not used).

A second fundamental principle is "ex nihilo nihil fit": nothing comes from nothing. Life does not arise by magic: it must come from properties of matter. Ditto for cognition. Ditto for consciousness. Many schemes have been proposed to explain how life or consciousness may be "created" from inanimate and unconscious matter, how a completely new property can arise from other properties. I don't believe this is the case. Both life and consciousness are ultimately natural phenomena that originate from other natural phenomena, just like television programs and the motion of stars.

I believe that the substance of the brain and the substance of consciousness are the same. Brain processes and thoughts arise from different properties of the same matter, just like a piece of matter exhibits both gravitational and electrical features. The feature that gives rise to consciousness is therefore present in every particle of the universe, just like the features that give rise to electricity and gravity.

Cognition is a feature of all matter, whether living or not: degrees of remembering, learning, etc. are ubiquitous in all natural systems. If you bend a piece of paper several times, it will tend to stay bent. That is equivalent to our memory memorizing something. If you leave it alone, the piece of paper will tend to resume its flat position. That is equivalent to our memory forgetting some information that is no longer used.

The issue, therefore, is not of what is conscious and what is not, of what is cognitive and what is not: the issue is the "degree" to which a system is conscious or cognitive. My degree of consciousness and of cognition are different from those of a stone, of a cat, of a tomato plant.

The explanation of consciousness does require a conceptual revolution in Science, specifically the introduction of a new feature of matter, which must be present even in the most fundamental building blocks of the universe. I believe that proto-consciousness is pervasive. Every piece of matter, down to the elementary constituents, is proto-conscious. The reason we "feel" is that each atom of our body "feels" (to some extent). Consciousness was there from the beginning. It is not created inside the brain by some magic process. Each neuron, and each atom of each neuron, is "proto-conscious". And each atom of every object is proto-conscious. The reason we are conscious is similar to the reason that some bodies are electrical conductors: each single particle of the universe has an electrical charge, and in some configurations that property yields conductivity. By the same token, each single particle of the universe has a proto-conscious quality, and in some configurations (for example, the human brain) that property yields consciousness.

My explanation of where our mind comes from goes like this. If consciousness is ubiquitous in nature, then it is not difficult to accept the idea that it was there, in some primitive form, since the very beginnings of life, and that it evolved with life. It became more and more complex as organisms became more and more complex. Early hominids were conscious, and their consciousness, while much more sophisticated than the consciousness of bacteria, was still rather basic, probably limited to fear, pain, pleasure, etc. Early hominids had a way to express through sounds their emotions of fear and pain and pleasure.

Consciousness was a skill that helped in natural selection. Minds were always busy thinking in very basic terms about survival, e.g. about how to avoid danger and how to create opportunities for food.

What set hominids apart from other mammals was the ability to manufacture tools. We can walk and we can use our hands in ways that no other animal can. The use of tools (weapons, clothes, houses, fire) relieved us from a lot of the daily processing that animals use their minds for. Our minds could afford to be "lazy". Instead of constantly monitoring the environment for preys and predators, our minds could afford to become "lazy". Out of that laziness modern consciousness was born. As the human mind had fewer and fewer practical chores, it could afford to do its own "gymnastics", rehearsing emotions, and constructing more and more complex ones. As more complex emotions helped cope with life, individuals who could generate and deal with them were rewarded by natural selection. Emotions underwent a Darwinian evolution of their own. That process is still occurring today.

Most animals cannot afford to spend much time philosophizing: their minds are constantly working to help them survive in their environment. Since tools were doing most of the job for us, our minds could afford the luxury of philosophizing, which is really mental gymnastics (to keep the mind in good shape).

In turn, this led to more and more efficient tools, to more and more mental gymnastics. As emotions grew more complex, sounds to express them grew more complex. It is not true that other animals cannot produce complex sounds. They have the sounds that express the emotions they feel. Human language developed to express more and more complex emotions. The quantity and quality of sounds kept increasing. Language trailed consciousness.

Ideas, or "memes", also underwent Darwinian evolution, spreading like virus from mind to mind, and continuously changing to adapt to new degrees of consciousness.

The history of consciousness is the history of the parallel and interacting evolution of: tools, language, memes, emotions and the brain itself. Each evolved and fostered the evolution of the others. The co-evolution of these "components" led to our current mental life.

This process continues today, and will continue for as long as tools allow more time for our minds to think. The software engineer daughter of a miner is "more" conscious than her father. And her father was more conscious than her ancestor who was, say, a medieval slave.

Consciousness is a product of having nothing better to do with our brain.

I also believe that the solution to the mystery of consciousness lies in a fundamental flaw of Physics. The two great theories of the universe that we have today, Quantum Physics and Relativity Theory, are incompatible. They both have an impressive record of achievements, but they are incompatible. One or both must go. I believe that once we replace them with one unified theory that is equally successful in explaining the cosmological realm and the subatomic realm, consciousness will be revealed to be a trivial consequence of the nature of the world. And I believe that this unified theory will be a "Theory of the Observer", not a theory of matter (as Physics has traditionally been). But that's material for another book.

Alas, in order to understand how I reached these conclusions you have to read the whole book. Once you have all the facts and all the theories, you can decide by yourself.

And let me clarify one more time: the vast majority of the book is not about my personal beliefs, but about the findings and the theories of specialists who know a lot more than me. In a sense, the goal of this book is to educate the reader to the point that the reader can work out her own theory of consciousness, cognition and life.

What do you think is the meaning of matter?

Summarizing What Is Summarizable

We don't really know why that particular piece of matter, the brain, yields consciousness. After all, the physical substance of the brain is made of the same elements found in all animals (carbon, hydrogen, oxygen, sodium, nitrogen, phosphorous, iron, calcium, potassium).

I have proposed a way out of this dilemma: to assume that a fundamental property of matter, of all matter, allows for the rise of consciousness when matter is organized in a particular manner. If consciousness is somehow present in each particle of the universe, then we don't need to explain the gap: there is no gap, just like there is no gap between electricity and gravity, they are simply different aspects of matter, which originate from different properties of fundamental particles.

During our long excursion in the maze of unsolved scientific problems, we progressively reduced much of our behavior (from common sense to emotions, from dreams to intelligence) and even life itself to a more and more mechanical process of interaction with the universe. The "Darwinian" theme kept coming back over and over again: we are the way we are not because somebody designed us that way but because a fundamental "force" of the universe allows for the survival of only those things that "fit" with the rest of the universe. The behavior of body and "mind" is actually very similar, once one takes this "Darwinian" perspective: both bodily organs and mental phenomena are bound to be what they are because they are useful for our survival, and both were shaped by external forces.

We are but small cogs in the gigantic machine of the universe. We don't live, we survive.

Ultimately, we don't live, we are lived.

Even worse: we don't think, we are thought.

We had to go through a cathartic reassessment of our role in the world (the Earth is not the center of the universe; man was not created by God; man is not the dominant species; the brain is just an organ; experience molds the brain) before being in a position to find what is truly unique about the human experience.

The final mystery is what this is all about. If we are but cogs in a huge machine, if our "minds" and bodies are but small machines that are part of a much bigger machine, what is this huge machine for?

The very meaning of life, the fundamental "why" of Science, can be rephrased as follows. Both the mind and the universe are machines that are computing something.

What?

A Word on Miracles

When I discuss Einstein's theory of Relativity with people who have no scientific background, my words are invariably met with a degree of skepticism, notwithstanding Hiroshima and all the other tangible proofs of its validity. On the other hand, mention a "miracle" to the same folks, and they will believe in it (to some degree) on the spot. By "miracle" I mean anything that eludes scientific proof, from telepathy to religious beliefs.

What is it that makes miracles so convincing, even if, over and over again, "miraculous" events and phenomena of the past have been explained rationally by science and proved nothing outside of the ordinary?

Miracles tend to occur only in places where there are no reliable news agencies and never in front of a video-camera. And they cannot be performed in front of scientists, but only in front of audiences that know little about Physics and Chemistry.

Not a single case of telepathy or levitation has been documented scientifically (the way one documents a flue outbreak or a car accident). Yet, most people believe they occur. The same people are often skeptic about Einstein's Relativity, no matter how many experiments have confirmed it.

Why does the human mind prefer miracles over science?

Human brains do not like scientific formulas, they like legends, stories, metaphors. The moment a scientist produces a rational explanation for a purported miracle, people need a new myth to replace it. Newton's and Einstein's brains were working unlike the brains of most people in that they were looking for rational theories to supplant myths.

The ordinary brain needs miracles. Miracles are hallucinations, random hallucinations, that have fed the brain since ancient times. Most brains don't know how to work without those hallucinations. Most brains cannot work in a purely logical manner, most brains need that irrational element.

Our brain is still very much dependent on the hallucinating voices that come from nowhere, have no explanation and prompt us to experience new behavior.

The Newtons and Einsteins were geniuses because they achieved a higher level of consciousness in which every miracle is questioned and no myth is ever generated. In those brains, the hallucinating voices have been reduced to only one, which expresses their quest for rationality. The chaotic noise of the hallucinating voices has been replaced by the ordered silence of logical thinking.

Not An Epilogue
People often ask me: "Do you believe in God?"
I reply: "Do you believe in zestykistirism?"
They say: "What's that?"
Precisely: "What is God?"

First of all, people should define what they mean by "God". As people struggle to define it, they often end up with such a vague definition that pretty much anything in the cosmos qualifies. The moment they try to be more specific, they fall into all sorts of traps that negate the very essence of the "God" that they are trying to define.

Sometimes, in order to avoid the logic around their own definitions, people claim that our minds cannot grasp the concept of God: if our minds cannot grasp the concept of the divine, then we have no clue what this God expects from us, thus the question is pointless.

The gods we worship (the Christian god, the Jewish god, the Islamic god, etc.) were invented by humans. They are reflections of humanity. They are as "human" as us. We assign them the moral values that we "value", the values that our mind can grasp. If superbeings exist, they are unlikely to be like anything we can envision, and such a superbeing could even be upset that we worship all these "human" gods.

(I also confess a deep antipathy for monotheism: I have always found the ancient polytheistic religions to be much more "civilized", i.e. tolerant and rational, than the tyrannical and dogmatic monotheistic religions of the Jews, the Christians and the Muslims. Mecca before Mohammed, where all "idols" were worshiped, is my ideal. Mecca after Mohammed, when all "other" idols became taboo, is my nightmare).

Thus the question about our "souls" and what happens to them after death is also meaningless. For the sake of arguing, one could ask what qualifies as a "soul" and what does not. The majority of religious people would probably reply that only humans qualify. They would claim that dogs have no soul, and worms have no soul, and spiders have no soul: these species do not go to paradise or hell, they simply disintegrate in the soil. Ditto for all the species that preceded Homo Sapiens Sapiens. The evidence? They can't speak, build cities, write books. One could ask: what about superior species? What if a new species appeared, that looks like a turtle but that is capable of doing everything we humans do (writing books, building cities and so forth) plus many other things that we humans can't do (reading minds or seeing in four dimensions or levitating). Would this species qualify for eternal life in paradise? If yes, then we reach the rather unpleasant conclusion that the human species is the "lowest" form of soul: anything inferior to a human being does not have a soul, while anything superior to a human being has a soul. If not, we reach the rather bizarre conclusion that, in order to have a soul, a being must be exactly like a human being, no less and no more: give or take a cognitive faculty and the being becomes a soul-less being. Either answer is unconvincing. The logical answer would be that every form of life is just that: a form of life. There are degrees of consciousness, emotion, feelings, and even of being alive. Some living beings are capable of things that other living beings are not capable of. And in the future new forms of life may appear, capable of actions that we are not capable of.

A Word On Mysticism
Sometimes the final proof of the existence of some divine force is ascribed to states of ecstasy, whether achieved via collective hysteria or some kind of drugs. What collective hysteria and hallucinogenic drugs have in common is that they, basically, paralyze parts of the brain, so that your perception is reduced (ironically, the opposite of what some psychedelic gurus used to claim) and your ability to perform reasoning is even further reduced. In other words, they make you dumber. As you become dumber, you have visions of God. Somehow the fact that one has to surrender her/his brain in order to grasp the existence of God makes me think that God is a remnant of a pre-human state of existence, one in which humans were not "sapiens", i.e. did not think the way they think today. I get the feeling that one can grasp the sense of God only if s/he returns to

the state of mind of primitive hominids. Perhaps all apes believe in God, because their brain is permanently in the state in which religions and psychedelic gurus want our brains to be. Perhaps all lower mammals, that do not have the neo-cortex and our cognitive faculties, "feel" God. Religion could simply be an evolutionary leftover, governed by the older part of the brain and superseded by reason whenever we let the newer part of the brain take over.

What is unique about humans is the elaborate rituals (involving dancing, chanting, drumming, funeral processions, masses for the dead, decorated tombs, etc) that they built around their ecstatic experiences. And the way that humans rationalized their ecstatic experiences into three fundamental "religious" beliefs: 1. That gods created the universe; 2. That humans are entitled to an afterlife; 3. That pleasing the gods will make the afterlife more pleasant and even eternal. Is this due to the interaction of the new rational brain with the older irrational brain? Is this the rational brain trying to make sense of the irrational brain?

Humans seem to live under the control of two genetic programs that operate in different directions. The newer genetic program, implemented as the neocortex, wants humans to be rational, while the older genetic program, implemented as the lower brain functions, wants humans to be irrational. The entire human civilization seems to be the outcome of this interplay between the rational and the irrational brain. Humans are maybe just a transitional species, between the irrational brain that has ruled the planet for millions of years and a new category of rational species, that will rule the planet for the next million years.

Emotions are very ancient and they are progressively disappearing.

The real appeal of studying religion for a scholar might be that it represents an earlier form of "thinking" in evolution.

What is probably unique to humans is that we are not only mystical but believe in supernatural beings. Human brains are equipped with the faculty to create a "theory of mind". We can speculate on what another person is thinking, and in many case we get it right. That faculty is due to our brain's ability to expect and recognize patterns of behavior. We then apply the same logic to "animate" objects, for example a tree that is swinging and making a sound because of the wind, or a thunder, or the moon. We assign them a mental life. (As we grow up we learn, or are taught, that they are not sentient beings). That is the origin of the spirits that populated Nature in primitive societies. Our brain naturally tends to explain a pattern of movement as caused by a mind. We recognize things not as objects but as subjects. When we see the pattern but don't see the "thing", we assign minds to invisible subjects, such as gods. Monotheism is the belief that the pattern of all patterns has a mind, is a subject.

What Does It All Mean?

What does it all mean? Is there an afterlife? I have a feeling that a) we cannot comprehend it, and b) the questions are not framed correctly. We cannot comprehend it because some things are beyond our cognitive closure, just like a snake cannot see in three dimensions and we can't see some colors of the spectrum. And the questions are not framed correctly because they refer to objects that are either not defined or improperly defined. Sometimes the logic is also odd: we are concerned about the afterlife, but seem indifferent to the "pre-life". We are scared that we will never exist again, but we are not scared that we never existed before. The eternity "after" our death terrifies us, but the eternity "before" our birth does not terrify us. Inevitably, one feels that our terror is programmed in our instinct for survival. When i wonder about the afterlife, i am just a machine programmed to avoid death.

It gets even worse if i meditate. When i meditate, i do not find peace at all: the fear of death increases, not decreases. But the more i meditate the more i realize that i also fear eternal life. It would be equally terrifying to live forever and ever and ever. In other words, i fear any non-human condition. I am a human being, programmed to live the finite life, and to desire the finite life, of a human being.

I do not envy an angel's eternal static life. I do not envy an angel because i am not controlled by an angel's genome. I am controlled by the human genome.

If i really had to imagine an afterlife, i wouldn't want to just be an eternal spectator on another dimension. I would hope that each of us can become a god herself or himself, and create her or his own universe. That would somewhat alleviate the boredom of eternity.

Extra-Terrestrial Life?

Is there extraterrestrial life? We first have to agree on what "life" means. If we mean "Earthly life" (the thing we call "life" here on Earth), then how much of it is enough to qualify as "life"? What thing between the fundamental constituents of life and a human being is enough to be considered "life"? Depending where one places the border, the chances of "life" existing somewhere else change dramatically. Do the constituents of Earthly life exist somewhere else? Absolutely. Do humans exist somewhere else? I do not think so, because it takes a very similar planet located near a very similar star at a very similar distance with a very similar geological history and a very similar history of astral encounters for billions of years to lead to human beings. (Calculations of the probability of Earthly life occurring on other planets routinely overlook the billions of cosmic events that shape the history of a planet).

If "life" does not have to be the kind that we find on this planet, then the question is what qualifies as "life" that is not made of the constituents of Earthly life (aminoacids and proteins). Since just about everything in the universe grows and decays to some extent (from rocks to Earthly life), just about everything can be squeezed into a broad definition of "life". Is a rock alive? Why not? It does change over time by interacting with the environment. It does multiply when it breaks into many pieces. It is just that Earthly animals do the same thing in a way that to us, Earthly animals, appears more complex.

The behaviorist approach is to define "life" as something that we would recognize as life because we can communicate with it to some extent. We cannot speak to a cat, but the cat reacts to our actions. That reaction tells us that the cat is "alive" (as opposed to a rock, that does not react and therefore we don't consider alive without any need for a biological definition of "life").

The problem is that communication too is a vague concept. Everything, ultimately, interacts with everything else. What degree of interaction is enough to qualify as communication?

My feeling is that we are using an Earthly term for non-Earthly processes when we should simply use a different name. Earthly life is just one of the many processes that can be found on planet Earth. There are many other processes, most of which have no name because we are not interested in naming them. For example, the way rocks decay on planet Earth is probably unique, but we have no name for the process of rock decay on planet Earth. You and I are not rocks, so we have no interest in giving this process a name.

Other planets simply host different kinds of processes. I feel that it is misleading to try to impose the term "life" (a term invented for an Earthly process) on whatever non-Earthly process. Why not just give it a different name?

How Will It End? The Future Evolution Of Humans – Part I

I am frequently asked about the future, about what is going to happen to humans in the future. Most futurologists simply reply that humans will self-destroy and will be replaced by a better race. That is a nice way to avoid figuring out what our civilization will be like millions of years from now. I do not believe that we will self-destroy, but i do believe that we will be replaced by newer and newer "races", descendants of our race (or, better, species).

In the short term i believe that humans will evolve into more and more conscious beings. I see this as a consequence of interaction with other humans and with the cultural specimen that other humans leave behind. The more we think the more... we think. It is a vicious loop that started when humans had the first philosophical conversation. We are more and more conscious of the human condition, of our (minuscule) role in the universe and of death. I see this trend continuing to degrees that today we can't even imagine. Homo Conscious is the successor of Homo Sapiens.

At the same time, i believe that some kind of regression in civilization will occur as a consequence of the stabilization and possible reduction of human population. Basically, i think that civilization was a result of population explosion, a self-sustaining positive feedback: as humans multiplied, each individual had to content herself with a smaller and smaller territory. Civilization was the art of making more of less. As an individual's territory shrank, he had to come up with more efficient means to provide for himself and his family. I believe that this is what civilization is all about. This process has been going for thousands of years, as the territory of an individual shrank from entire forests to an apartment in a high-rise building (mostly not even owned, but only rented). People work because they can no longer rely on a natural profusion of food and materials: competition with other humans has been making those resources scarcer and scarcer. This has triggered human creativity, and caused the advent of science and technology, and everything else that we call "civilization".

Once this process is reversed, I believe that civilization will reverse too. An abundance of natural resources would automatically defuse human creativity.

Thus i foresee a near-term future (a few centuries from now) in which humans will become more and more conscious while being more and more "savage".

How Will It End? The Future Evolution Of Humans – Part II
In a few centuries or millennia, I think that humans will have to face another "intelligent" species. I am not referring to extraterrestrial life (if it exists, it is just so unlikely to interact with Earthly life until we start traveling very long distances). I am referring to descendants of species that exist right now on Earth. I believe that some of them will eventually become "intelligent" enough that humans will have to consider them as equal as, say, very dumb people (ubiquitous in every continent, and often even elected to office). I believe that other animal species will eventually evolve the ability to create civilizations, more or less similar to the ones created by humans (minus human language, of course, which will remain a

human peculiarity). I have a hunch that birds (rather than mammals) are the best candidates for such a breakthrough: they live in large societies and they travel long distances. Somehow, i feel that these are prerequisites to the emergence of a higher form of consciousness.

I don't think humans will ever have to share their world with aliens, but i do think that humans will eventually have to share their world with other Earthly species as capable as humans of building civilizations.

How Will It End? The Future Evolution Of Humans – Part III

We can argue forever about what happened during the evolution of the human brain, but what happened to the human body seems pretty clear: it got bigger, and it is still getting bigger, generation after generation. Of course, diet matters. But diet mattered also millions of years ago. It is part of the story of evolution.

We tend to focus on the evolution of the brain, not on the evolution of the body, but i think that the brain evolves to serve the body. Therefore i am interested in finding out the ways in which the human body can evolve.

At the same time, i have always been intrigued by the fact that this human race, so good at expanding beyond its original environment, has scant chances of expanding beyond our neighborhood (the Solar System), of ever exploring the universe: we are too small and our lives are too short for us to conquer unlimited space and time. It is difficult to believe that any kind of scientific progress will allow such small and brief beings to reach any other galaxy.

One day I finally realized the simplest way that humans could colonize the universe: if they evolved into bigger beings, their task of explorers would be much easier, just like an elephant can more easily travel long distances than an ant can. It is not a question of brains, it is simply a matter of size. If humans evolve into a race of giants (giants the size of planets), then they will be able to explore space by simply "taking a walk". These giants, who will eat planets and drink comets (and live millions of years), will be able to "hike" from galaxy to galaxy. If they invent their own transportation system, they might be able to cover distances of billions of light years. (Is this impossible according to Relativity? It is difficult to tell if Relativity, and any other human science, is a true representation of the universe as it is, or just the best representation of the universe as today's humans see it). Of course, this is not something likely to happen in the next few millennia.

But i do think that the problem of space exploration will be solved by biological evolution, not by technological evolution. I think the former has higher chances of succeeding than the latter. Technology will reach a point where improvement will come at a slower and slower pace, unless it is matched by significant biological evolution. On the other hand, biological evolution may speed up in response to climate change or some galactic event, and the current trend towards bigger bodies may get amplified tenfold. Over millions of generations, I can visualize descendants of the human race that managed to survive all possible catastrophes and got so big (millions of times bigger) that they have become cosmic objects and roam the universe the same way that we roam the Earth.

They will know about us, the same way we know about the fossils of species that preceded us. They will know of our civilizations, or, better, the fossils of civilizations that preceded theirs. They will study us the same way we study a fossil. No matter how much information we produce in no matter how many different media, our fossils will look devoid of crucial information that is relevant to them for the simple reason that we cannot conceive of the information that will be relevant millions of years from now.

But all of this is a big "if": if our form of life survives long enough. Unfortunately, it is also likely that life on Earth will eventually be destroyed by a galactic catastrophe and nobody will ever know that we existed.

How Did It Start?

It is popular to claim that, thanks to the advent of the human race, the universe has somehow acquired awareness of itself. After billions of years of blind and dumb evolution, the universe finally created us humans, who are endowed with the faculties of consciousness, and therefore in us the universe has created a part of itself that is aware of the whole.

That's a human-centric view that is difficult to prove. It could well be that the universe has always been aware of itself, and we humans are just the current manifestation of that self-awareness (one of the current ones).

And it could well be that we are "not" aware of the universe but just of our little niche.

Prologue

I also envision another possible future, but it is harder to explain.

If we are part of the universe that we study, and we are becoming more and more conscious of what the universe is and does, and more and more of the universe is becoming "us", then maybe the "what" is a process of self-discovery: the universe is going through a pain-staking process of self-discovery. As conscious life triumphs over unconscious matter, the entire universe will become conscious, aware of itself.

In a sense, we are just waking up, as if after a long sleep, and slowly beginning to realize where we are and who we are.

Thankfully we cannot understand why. Thankfully we cannot understand the universe.

Alphabetical Index of Names

Allen, Woody : 22
Arbib, Michael : 48
Armstrong, David : 49
Aserinsky, Eugene : 7
Baars, Bernard : 102
Baird, Bill : 98
Bateson, Gregory : 87
Bell, John : 124
Benington, Joel : 18
Bergson, Henri : 19
Berkeley, George : 128
Bernard, Claude : 163
Birdwhistell, Ray : 26
Blackmore, Susan : 106
Blechner, Mark : 19
Bohm, David : 125, 126, 127, 131, 132, 140
Bohr, Niels : 140
Borb,ly, Alexander : 7
Bose, Satyendranath : 122
Bruner, Jerome : 149
Buck, Ross : 28
Buonomano, Dean : 111
Cairns, John : 123
CairnsSmith, Graham : 60
Calvin, William : 19, 51, 107
Campbell, Donald : 148
Campbell, Joseph : 16, 19, 21, 109
Campbell, Scott : 8
Canamero, Lola : 40
Carlson, Richard : 153
Cauvin, Jacques : 56
Chalmers, David : 135, 139
Chomsky, Noam : 10, 20, 21, 64
Churchland, Paul : 98
Clifford, William : 140
Conrad, Michael : 134
Cornell, Eric : 122
Crick, Francis : 12, 94
Culbertson, James : 125
Damasio, Antonio : 32, 35, 89
Darwin, Charles : 7, 40, 85

Dawkins, Richard : 80
Deacon, Terrence : 48
Decety, Jean : 30
Dement, William : 7, 18
Dennett, Daniel : 51, 60, 63, 100, 105, 151, 153
Descartes, Rene : 119
Descartes, Rene : 83, 118, 139
DeSousa, Ronald : 27
Deutsch, David : 80
Dirac, Paul : 120
Donald, Merlin : 57, 63
Dunbar, Robin : 64
Eccles, John : 52, 88, 121, 167
Edelman, Gerald : 91
Edgar, Dale : 18
Einstein, Albert : 120, 122
Eisenberg, Nancy : 30
Ekman, Paul : 40
Festinger, Leon : 104
Flanagan, Owen : 16, 88
Fodor, Jerry : 54
Fraser, Julius : 110
Freeman, Walter : 94, 113, 151
Freud, Sigmund : 5
Frijda, Nico : 29
Froehlich, Herbert : 122
Fu, YingHui : 8
Gabor, Dennis : 132
Gardenfors, Peter : 61
Gazzaniga, Michael : 51, 100, 103, 151
Geertz, Clifford : 40
Gibson, James : 87
Goldie, Peter : 40
Goldstein, Kurt : 34
Goswami, Amit : 129
Gray, Charles : 96
Greenfield, Susan : 100
Griffin, David : 140
Griffiths, Paul : 40
Grimm, Rudolf : 122

Grof, Staninslav : 87
Haeckel, Ernst : 140
Haldane, John : 80, 140
Hameroff, Stuart : 64, 134
Harth, Erich : 72, 100
Hartshorne, Charles : 140
Hebb, Donald : 108
Heidegger, Martin : 30
Heisenberg, Werner : 119, 127
Heller, Craig : 18
Herbert, Nick : 124
Hinton, Geoffrey : 8
Hobson, Allan : 8, 12, 26, 109
Hoffman, Martin : 30
Hofstadter, Douglas : 159
Hoppensteadt, Frank : 99
Humphrey, Nicholas : 48, 87, 91
Hyden, Holger : 17
IngeboBarth, Denise : 108
Izhikevich, Eugene : 99
Jackendoff, Ray : 88
Jackson, Frank : 75
Jackson, Hughlings : 7
Jackson, Philip : 30
James, William : 19, 24, 29, 33, 35, 84, 88, 106, 127, 141
Jauregui, Jose : 30
Jaynes, Julian : 49
Jerison, Harry : 52
Jin, Deborah : 122
Jouvet, Michel : 7
Jung, Carl : 10, 19, 107, 132
Kant, Immanuel : 79, 87
KarmiloffSmith, Annette : 54, 58
Kinsbourne, Marcel : 108
Kleitman, Nathaniel : 7
Klopf, Harry : 28
Koch, Christof : 94
Kotre, John : 152
Laird, James : 35
Lang, Peter : 36
Lange, Paul : 17
Lashley, Karl : 77
Laszlo, Erwin : 131
Lazarus, Richard : 29
Ledoux, Joseph : 25

Lewis, Clarence : 74
Libet, Benjamin : 104, 150
Llinas, Rodolfo : 18, 97
Locke, John : 87
Lockwood, Michael : 121
Lotka, Alfred : 119
Lotze, Hermann : 140
Luz, Catherine : 40
Lycan, William : 87
Mach, Ernst : 140
MacLean, Paul : 40, 59
MacPhail, Euan : 56
Malsburg, Christoph : 95
Mandler, George : 29
Margulis, Lynn : 65
Marshall, Ian : 123
Mayer, John : 29
McCarley, Robert : 8
McGinn, Colin : 77, 79, 83
McNamara, Patrick : 19
Mead, George : 48, 85
Minsky, Marvin : 21
Mithen, Steven : 53
Nagel, Thomas : 77, 84, 140
Neisser, Ulric : 150, 152, 153
Newton, Isaac : 84, 118
Norretranders, Tor : 111
Nunez, Paul : 98, 113, 151
Oakley, Kenneth : 52
Olson, Eric : 159
Ornstein, Robert : 104, 150
Panksepp, Jaak : 25
Parfit, Derek : 151, 155
Peirce, Charles : 127
Peirce, Charles : 141
Penrose, Roger : 134, 167
Piaget, Jean : 58
Picard, Rosalind : 40
Plutchik, Robert : 40
Popper, Karl : 47, 106, 141, 154
Porges, Stephen : 59
Pribram, Karl : 132
Pylkkanen, Paavo : 127
Rolls, Edmund : 38
Russell, Bertrand : 76, 79, 121
Ryle, Gilbert : 160

Salovey, Peter : 29
Saper, Clifford : 8
Sartre, JeanPaul : 30
Scaruffi, Piero : 1, 2, 66, 141, 172
Schroedinger, Erwin : 65
Searle, John : 83, 153
Shannon, Claude : 97
Siapas, Thanos : 111
Simon, Herbert : 27
Singer, Wolf : 94
Sirag, Saul : 121
Sloman, Aaron : 27
Snell, Bruno : 49
Snyder, Fred : 15
Solms, Mark : 18
Solomon, Robert : 30
Sperry, Roger : 103, 148
Stapp, Henry : 127
Stenning, Keith : 36
Stevens, Anthony : 15
Teilhard, Pierre : 132, 140
Thomas, Lewis : 159
Tiller, William : 132
Tobler, Irene : 8
Tomasello, Michael : 59
Tononi, Giulio : 97
Turing, Alan : 96
Umezawa, Hiroomi : 133
Varela, Francisco : 99
VonHelmholtz, Hermann : 85
VonNeumann, John : 127
Vygotsky, Lev : 58
Walker, Evan : 119
Wegner, Daniel : 150
Whitehead, Alfred : 140
Wieman, Carl : 122
Wiener, Norbert : 148
Wigner, Eugene : 120, 127
Wigner, Paul : 129
Wilson, EdwardOsborne : 30
Winson, Jonathan : 10, 109
Wittgenstein, Ludwig : 36
Wolf, Fred : 130
Yasue, Kunjo : 133
Zajonc, Robert : 24
Zhang, Jie : 18

Ziehen, Theodore : 141
Zohar, Danah : 124

www.ingramcontent.com/pod-product-compliance
Lightning Source LLC
Chambersburg PA
CBHW051650170526
45167CB00001B/410